General-Purpose
Graphics Processor Architectures

Synthesis Lectures on Computer Architecture

Editor
Margaret Martonosi, *Princeton University*

Founding Editor Emeritus
Mark D. Hill, *University of Wisconsin, Madison*

Synthesis Lectures on Computer Architecture publishes 50- to 100-page publications on topics pertaining to the science and art of designing, analyzing, selecting and interconnecting hardware components to create computers that meet functional, performance and cost goals. The scope will largely follow the purview of premier computer architecture conferences, such as ISCA, HPCA, MICRO, and ASPLOS.

General-Purpose Graphics Processor Architectures

Tor M. Aamodt, Wilson Wai Lun Fung, and Timothy G. Rogers

ISBN: 978-3-031-00631-9   paperback
ISBN: 978-3-031-01759-9   ebook
ISBN: 978-3-031-00056-0   hardcover

DOI 10.1007/978-3-031-01759-9

A Publication in the Springer series
SYNTHESIS LECTURES ON COMPUTER ARCHITECTURE

Lecture #44
Series Editor: Margaret Martonosi, *Princeton University*
Founding Editor Emeritus: Mark D. Hill, *University of Wisconsin, Madison*
Series ISSN
Print 1935-3235   Electronic 1935-3243

General-Purpose
Graphics Processor Architectures

Tor M. Aamodt
University of British Columbia

Wilson Wai Lun Fung
Samsung Electronics

Timothy G. Rogers
Purdue University

SYNTHESIS LECTURES ON COMPUTER ARCHITECTURE #44

ABSTRACT

Originally developed to support video games, graphics processor units (GPUs) are now increasingly used for general-purpose (non-graphics) applications ranging from machine learning to mining of cryptographic currencies. GPUs can achieve improved performance and efficiency versus central processing units (CPUs) by dedicating a larger fraction of hardware resources to computation. In addition, their general-purpose programmability makes contemporary GPUs appealing to software developers in comparison to domain-specific accelerators. This book provides an introduction to those interested in studying the architecture of GPUs that support general-purpose computing. It collects together information currently only found among a wide range of disparate sources. The authors led development of the GPGPU-Sim simulator widely used in academic research on GPU architectures.

The first chapter of this book describes the basic hardware structure of GPUs and provides a brief overview of their history. Chapter 2 provides a summary of GPU programming models relevant to the rest of the book. Chapter 3 explores the architecture of GPU compute cores. Chapter 4 explores the architecture of the GPU memory system. After describing the architecture of existing systems, Chapters 3 and 4 provide an overview of related research. Chapter 5 summarizes cross-cutting research impacting both the compute core and memory system.

This book should provide a valuable resource for those wishing to understand the architecture of graphics processor units (GPUs) used for acceleration of general-purpose applications and to those who want to obtain an introduction to the rapidly growing body of research exploring how to improve the architecture of these GPUs.

KEYWORDS

GPGPU, computer architecture

Contents

Preface

This book is intended for those wishing to understand the architecture of graphics processor units (GPUs) and to obtain an introduction to the growing body of research exploring how to improve their design. It is assumed readers have a familiarity with computer architecture concepts such as pipelining and caches and are interested in undertaking research and/or development related to the architecture of GPUs. Such work tends to focus on trade-offs between different designs, and thus this book is written with a view to providing insights into such trade-offs so that the reader can avoid having to learn by trial and error what is already known to experienced designers.

To help achieve this, the book collects together into one resource many relevant bits of information currently found among a wide range of disparate sources such as patents, product documents, and research papers. It is our hope this will help reduce the time it takes for a student or practitioner just starting to do their own research to become productive.

While this book covers aspects of current GPU designs, it also attempts to "synthesize" published research. This is partly due to necessity, as very little has been said by vendors on the microarchitecture of specific GPU products. In describing a "baseline" GPGPU architecture, this book relies both upon published product descriptions (journal papers, whitepapers, manuals) and, in some cases, descriptions in patents. The details found in patents may differ substantially from the microarchitecture of actual products. In some cases, microbenchmark studies have clarified for researchers some details, but in others our baseline represents our "best guess" based upon publicly available information. Nonetheless, we believe this will be helpful as our focus is understanding architecture trade-offs that have already been studied or might be interesting to explore in future research.

Several portions of this book focus on summarizing the many recent research papers on the topic of improving GPU architectures. As this topic has grown significantly in popularity in recent years, there is too much to cover in this book. As such, we have had to make difficult choices about what to cover and what to leave out.

Tor M. Aamodt, Wilson Wai Lun Fung, and Timothy G. Rogers
April 2018

Acknowledgments

We would like to thank our families for their support while writing this book. Moreover, we thank our publisher, Michael Morgan and editor, Margaret Martonosi, for the extreme patience they have shown while this book came together. We also thank Carole-Jean Wu, Andreas Moshovos, Yash Ukidave, Aamir Raihan, and Amruth Sandhupatla for providing detailed feedback on early drafts of this book. Finally, we thank Mark Hill for sharing his thoughts on strategies for writing Synthesis Lectures and specific suggestions for this book.

Tor M. Aamodt, Wilson Wai Lun Fung, and Timothy G. Rogers
April 2018

CHAPTER 1

Introduction

This book explores the hardware design of graphics processor units (GPUs). GPUs were initially introduced to enable real-time rendering with a focus on video games. Today GPUs are found everywhere from inside smartphones, laptops, datacenters, and all the way to supercomputers. Indeed, an analysis of the Apple A8 application processor shows that it devotes more die area to its integrated GPU than to central processor unit (CPU) cores [A8H]. The demand for ever more realistic graphics rendering was the initial driver of innovation for GPUs [Montrym and Moreton, 2005]. While graphics acceleration continues to be their primary purpose, GPUs increasingly support non-graphics computing. One prominent example of this receiving attention today is the growing use GPUs to develop and deploying machine learning systems [NVIDIA Corp., 2017]. Thus, the emphasis of this book is on features relevant to improving the performance and energy efficiency of non-graphics applications.

This introductory chapter provides a brief overview of GPUs. We start in Section 1.1 by considering the motivation for the broader category of computation accelerators to understand how GPUs compare to other options. Then, in Section 1.2, we provide a quick overview of contemporary GPU hardware. Finally, Section 1.4 provides a roadmap to the rest of this book.

1.1 THE LANDSCAPE OF COMPUTATION ACCELERATORS

For many decades, succeeding generations of computing systems showed exponential increasing performance per dollar. The underlying cause was a combination of reduced transistor sizes, improvements in hardware architecture, improvements in compiler technology, and algorithms. By some estimates half of those performance gains were due to reductions in transistor size that lead to devices that operate faster [Hennessy and Patterson, 2011]. However, since about 2005, the scaling of transistors has failed to follow the classical rules now known as Dennard Scaling [Dennard et al., 1974]. One key consequence is that clock frequencies now improve much more slowly as devices become smaller. To improve performance requires finding more efficient hardware architectures.

By exploiting hardware specialization it is possible to improve energy efficiency by as much as 500× [Hameed et al., 2010]. As shown by Hameed et al., there are several key aspects to attaining such gains in efficiency. Moving to vector hardware, such as that found in GPUs, yields about a 10× gain in efficiency by eliminating overheads of instruction processing. A large part of the remaining gains of hardware specialization are a result of minimizing data movement which can be achieved by introducing complex operations that perform multiple arithmetic operations while avoiding accesses to large memory arrays such as register files.

A key challenge for computer architects today is finding better ways to balance the gains in efficiency that can be obtained by using specialized hardware with the need for flexibility required to support a wide range of programs. In the absence of architectures only algorithms that can be used for a large number of applications will run efficiently. An emerging example is hardware specialized for supporting deep neural networks such as Google's Tensor Processing Unit [Jouppi et al., 2017]. While machine learning appears likely to occupy a very large fraction of computing hardware resources, and these may migrate to specialized hardware, we argue there will remain a need for efficiently supporting computation expressed as software written in traditional programming languages.

One reason for the strong interest in GPU computing outside of the use of GPUs for machine learning is that modern GPUs support a Turing Complete programming model. By Turing Complete, we mean that any computation can be run given enough time and memory. Relative to special-purpose accelerators, modern GPUs are flexible. For software that can make full use of GPU hardware, GPUs can be an order of magnitude more efficient than CPUs [Lee et al., 2010]. This combination of flexibility and efficiency is highly desirable. As a consequence many of the top supercomputers, both in terms of peak performance and energy efficiency now employ GPUs [top]. Over succeeding generations of products, GPU manufacturer's have refined the GPU architecture and programming model to increase flexibility while simultaneously improving energy efficiency.

1.2   GPU HARDWARE BASICS

Often those encountering GPUs for the first time ask whether they might eventually replace CPUs entirely. This seems unlikely. In present systems GPUs are not stand-alone computing devices. Rather, they are combined with a CPU either on a single chip or by inserting an add-in card containing only a GPU into a system containing a CPU. The CPU is responsible for initiating computation on the GPU and transferring data to and from the GPU. One reason for this division of labor between CPU and GPU is that the beginning and end of the computation typically require access to input/output (I/O) devices. While there are ongoing efforts to develop application programming interfaces (APIs) providing I/O services directly on the GPU, so far these all assume the existence of a nearby CPU [Kim et al., 2014, Silberstein et al., 2013]. These APIs function by providing convenient interfaces that hide the complexity of managing communication between the CPU and GPU rather than eliminating the need for a CPU entirely. Why not eliminate the CPU? The software used to access I/O devices and otherwise provide operating system services would appear to lack features, such as massive parallelism, that would make them suitable to run on the GPU. Thus, we start off by considering the interaction of the CPU and GPU.

An abstract diagram showing a typical system containing a CPU and GPU is shown in Figure 1.1. On the left is a typical discrete GPU setup including a bus connecting the CPU and GPU (e.g., PCIe) for architectures such as NVIDIA's Volta GPU, and on the right is a logical diagram of a typical integrated CPU and GPU such as AMD's Bristol Ridge APU or a mobile GPU. Notice that systems including discrete GPUs have separate DRAM memory spaces for the CPU (often called system memory) and the GPU (often called device memory). The DRAM technology used for these memories is often different (DDR for CPU vs. GDDR for GPU). The CPU DRAM is typically optimized for low latency access whereas the GPU DRAM is optimized for high throughput. In contrast, systems with integrated GPUs have a single DRAM memory space and therefore necessarily use the same memory technology. As integrated CPUs and GPUs are often found on low-power mobile devices the shared DRAM memory is often optimized for low power (e.g., LPDDR).

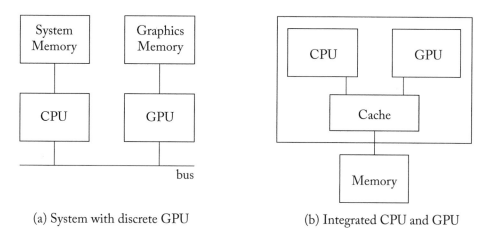

(a) System with discrete GPU   (b) Integrated CPU and GPU

Figure 1.1: GPU computing systems include CPUs.

A GPU computing application starts running on the CPU. Typically, the CPU portion of the application will allocate and initialize some data structures. On older discrete GPUs from both NVIDIA and AMD the CPU portion of the GPU Computing application typically allocates space for data structures in both CPU and GPU memory. For these GPUs, the CPU portion of the application must orchestrate the movement of data from CPU memory to GPU memory. More recent discrete GPUs (e.g., NVIDIA's Pascal architecture) have software and hardware support to automatically transfer data from CPU memory to GPU memory. This can be achieved by leveraging virtual memory support [Gelado et al., 2010], both on the CPU and GPU. NVIDIA calls this "unified memory." On systems in which the CPU and GPU are integrated onto the same chip and share the same memory, no programmer controlled copying from CPU memory to GPU memory is necessary. However, because CPUs and GPUs use caches and some of these caches may be private, there can be a cache-coherence problem, which hardware developers need to address [Power et al., 2013b].

At some point, the CPU must initiate computation on the GPU. In current systems this is done with the help of a driver running on the CPU. Before launching computation on the GPU, a GPU computing application specifies which code should run on the GPU. This code is commonly referred to as a kernel (more details in Chapter 2). At the same time the CPU portion of the GPU computing application also specifies how many threads should run and where these threads should look for input data. The kernel to run, number of threads, and data location are conveyed to the GPU hardware via the driver running on the CPU. The driver will translate the information and place it memory accessible by the GPU at a location where the GPU is configured to look for it. The driver then signals the GPU that it has new computations it should run.

A modern GPU is composed of many cores, as shown in Figure 1.2. NVIDIA calls these cores *streaming multiprocessors* and AMD calls them *compute units*. Each GPU core executes a single-instruction multiple-thread (SIMT) program corresponding to the kernel that has been launched to run on the GPU. Each core on a GPU can typically run on the order of a thousand threads. The threads executing on a single core can communicate through a scratchpad memory and synchronize using fast barrier operations. Each core also typically contains first-level instruction and data caches. These act as bandwidth filters to reduce the amount of traffic sent to lower levels of the memory system. The large number of threads running on a core are used to hide the latency to access memory when data is not found in the first-level caches.

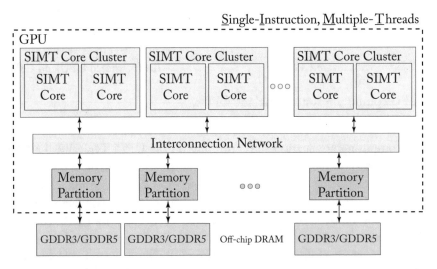

Figure 1.2: A generic modern GPU architecture.

To sustain high computation throughput it is necessary to balance high computational throughput with high memory bandwidth. This in turn requires parallelism in the memory system. In GPUs this parallelism is provided by including multiple memory channels. Often, each memory channel has associated with it a portion of last-level cache in a memory partition. The GPU cores and memory partitions are connected via an on-chip interconnection network such as a crossbar. Alternative organizations are possible. For example, the Intel Xeon Phi, which directly competes with GPUs in the supercomputing market, distributes the last-level cache with the cores.

GPUs can obtain improved performance per unit area vs. superscalar out-of-order CPUs on highly parallel workloads by dedicating a larger fraction of their die area to arithmetic logic units and correspondingly less area to control logic. To develop intuition into the tradeoffs between CPU and GPU architectures, Guz et al. [2009] developed an insightful analytical model showing how performance varies with number of threads. To keep their model simple, they assume a simple cache model in which threads do not share data and infinite off-chip memory bandwidth. Figure 1.3 which reproduces a figure from their paper, illustrates an interesting trade-off they found with their model. When a large cache is shared among a small number of threads (as is the case in multicore CPUs), performance increases with the number of threads. However, if the number of threads increases to the point that the cache cannot hold the entire working set, performance decreases. As the number of threads increases further, performance increases with the ability of multithreading to hide long off-chip latency. GPUs architectures are represented by the right-hand side of this figure. GPUs are designed to tolerate frequent cache misses by employing multithreading.

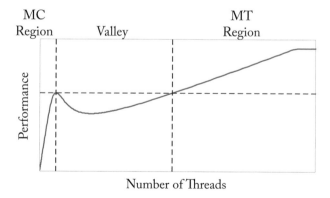

Figure 1.3: An analytical model-based analysis of the performance tradeoff between multicore (MC) CPU architectures and multithreaded (MT) architectures such as GPUs shows a "performance valley" may occur if the number of threads is insufficient to cover off-chip memory access latency (based on Figure 1 from Guz et al. [2009]).

With the end of Dennard Scaling [Horowitz et al., 2005], increasing energy efficiency has become a primary driver of innovation in computer architecture research. A key observation is that accessing large memory structures can consume as much or more energy as computation.

For example, Table 1.1 provides data on the energy for various operations in a 45 nm process technology [Han et al., 2016]. When proposing novel GPU architecture designs it is important to take energy consumption into account. To aid with this, recent GPGPU architecure simulators such as GPGPU-Sim [Bakhoda et al., 2009] incorporate energy models [Leng et al., 2013].

Table 1.1: Energy consumption of various operations for a 45 nm process technology (based on Table 1 in Han et al. [2016])

Operation	Energy [pJ]	Relative Cost
32 bit int ADD	0.1	1
32 bit float ADD	0.9	9
32 bit int MULT	3.1	31
32 bit float MULT	3.7	37
32 bit 32KB SRAM	5	50
32 bit DRAM	640	6400

1.3 A BRIEF HISTORY OF GPUS

This section briefly describes the history of graphics processing units. Computer graphics emerged in the 1960s with projects such as Ivan Sutherland's Sketchpad [Sutherland, 1963]. From its earliest days computer graphics have been integral to off-line rendering for animation in films and in parallel the development of real-time rendering for use in video games. Early video cards started with the IBM Monochrome Display Adapter (MDA) in 1981 which only supported text. Later, video cards introduced 2D and then 3D acceleration. In addition to video games 3D accelerators targeted computer-aided design. Early 3D graphics processors such as the NVIDIA GeForce 256 were relatively fixed-function. NVIDIA introduced programmability to the GPU in the form of vertex shaders [Lindholm et al., 2001] and pixel shaders in the GeForce 3 introduced in 2001. Researchers quickly learned how to implement linear algebra using these early GPUs by mapping matrix data into into textures and applying shaders [Krüger and Westermann, 2003] and academic work at mapping general-purpose computing onto GPUs such that the programmer did not need to know graphics soon followed [Buck et al., 2004]. These efforts inspired GPU manufacturers to directly support general-purpose computing in addition to graphics. The first commercial product to do so was the NVIDIA GeForce 8 Series. The GeForce 8 Series introduced several innovations including ability to write to arbitrary memory addresses from a shader and scratchpad memory to limit off-chip bandwidth, which had been lacking in earlier GPUs. The next innovation was enabling caching of read-write data with NVIDIA's Fermi architecture. Subsequent refinements include AMD's Fusion architecture which integrated CPU and GPU on the same die and dynamic parallelism that enables launching of threads from the GPU itself. Most recently, NVIDIA's Volta introduces features such as Tensor Cores that are targeted specifically at machine learning acceleration.

1.4   BOOK OUTLINE

The rest of this book is organized as follows.

When designing hardware it is important to consider the software that it will support. Thus, in Chapter 2, we provide a brief summary of the programming model, code development process, and compilation flow.

In Chapter 3, we explore the architecture of individual GPU cores that support execution of thousands of threads. We incrementally build up an increasingly detailed understanding of the trade-offs involved in supporting high throughput and a flexible programming model. The chapter finishes up by summarizing recent research related to the architecture of GPU cores to help quickly bring up to speed those new to the field.

In Chapter 4, we explore the memory system including both the first-level caches found within the GPU cores, and the internal organization of the memory partitions. It is important to understand the memory system of GPUs as computations that run on GPUs are often limited by off-chip memory bandwidth. The chapter finishes up by summarizing recent research related to GPU memory system architecture.

Finally, Chapter 5 gives an overview of additional research on GPU computing architectures that does not neatly fit into Chapter 3 or 4.

C H A P T E R   2

Programming Model

The goal of this chapter is to provide enough context about how GPUs are programmed for non-graphics computing so that those who have no prior experience with GPUs can follow the discussion in later chapters. We focus here on essential material, leaving more in-depth coverage to other references (e.g., [Kirk and Wen-Mei, 2016]). Many GPU computing benchmark suites exist which can be employed for architecture research. Learning how GPUs are programmed is relevant to a computer architect interested in GPU computing to gain a better understanding of the hardware/software interface but it becomes essential if you want to explore making changes to the hardware/software interface as part of your research. In the latter case, existing benchmarks may not exist and so will may need to be created, perhaps by modifying the source code of existing GPU computing applications. For example, research exploring the introduction of transactional memory (TM) on GPUs required this because current GPUs do not support TM (see Section 5.3).

Modern GPUs employ wide SIMD hardware to exploit the data-level parallel in GPU applications. Instead of exposing this SIMD hardware directly to the programmer, GPU computing APIs, such as CUDA and OpenCL, feature a MIMD-like programming model that allows the programmer to launch a large array of scalar threads onto the GPU. Each of these scalar threads can follow its unique execution path and may access arbitrary memory locations. At runtime, the GPU hardware executes groups of scalar threads, called *warps* (or *wavefronts* in AMD terminology), in lockstep on SIMD hardware to exploit their regularities and spatial localities. This execution model is called single-instruction, multiple-thread (SIMT) [Lindholm et al., 2008a, Nickolls and Reusch, 1993].

The rest of this chapter expands upon this discussion and is organized as follows. In Section 2.1, we explore the conceptual execution model used by recent GPU programming models and provide a concise summary of the execution model for typical GPUs released in the past decade. In Section 2.2, we explore the compilation process for GPU computing applications and take a brief look at GPU instruction set architectures.

2.1   EXECUTION MODEL

A GPU computing application starts execution on a CPU. For discrete GPUs, the CPU portion of the application will typically allocate memory for use in the computation on the GPU and then initiate transfer of input data into GPU memory, and finally launch a computational kernel on the GPU. For integrated GPUs only the last step is necessary. A computational kernel is composed of (typically) thousands of threads. Each thread executes the same program, but may follow a different control flow through that program depending upon the outcome of the computation. Below we consider this flow in detail using a specific code example written in CUDA. In the following section we look at the execution model at the assembly level. Our discussion does not dwell on performance aspects of GPU programming models. However, one interesting observation made by Seo et al. [2011] in the context of OpenCL (a programming model similar to CUDA which can be compiled to many architectures) is that code carefully optimized for one architecture (e.g., a GPU) may perform poorly on another (e.g., a CPU).

Figure 2.1 provides C code for a CPU implementation of the well-known operation *single-precision scalar value A times vector value X plus vector value Y*, known as SAXPY. SAXPY is part of the well-known Basic Linear Algebra Software (BLAS) library [Lawson et al., 1979], and is useful for implementing higher level matrix operations such as Gaussian elimination [McCool et al., 2012]. Given its simplicity and utility, it is often used as an example when teaching computer architecture [Hennessy and Patterson, 2011]. Figure 2.2 provides a corresponding CUDA version of SAXPY that splits execution across a CPU and GPU.

The example in Figure 2.2 demonstrates the abstraction provided by CUDA and related programming models (e.g., OpenCL [Kaeli et al., 2015]). The code begins execution with the function main(). To keep the example focused on details specific to computation on the GPU we omit details of allocating and initializing the arrays x and y. Next, the function saxpy_serial is called. This function takes as input arguments the number of elements in the vectors x and y in parameter n, the scalar value in parameter a, and pointers to arrays used to represent the vectors x and y. The function iterates over each element of the arrays x and y. In each iteration the code on line 4 reads the values x[i] and y[i] using the loop variable i, multiplies x[i] by a then adds y[i], and then updates x[i] with the result. For simplicity, we omit details of how the CPU uses the result of the function call.

Next, we consider a CUDA version of SAXPY. Similar to a traditional C or C++ program the code in Figure 2.2 begins execution by running the function main() on the CPU. Rather than walking through this code line by line, we will first highlight aspects specific to GPU execution.

Threads that execute on the GPU are part of a compute *kernel* specified by a function. In the CUDA version of SAXPY, shown in Figure 2.2, the CUDA keyword __global__ on line 1 indicates the kernel function saxpy will run on the GPU. In the example in Figure 2.2 we have parallelized the "for" loop from Figure 2.1. Specifically, each iteration of the "for" loop on line 4 in the original CPU-only C code from Figure 2.1 is translated into an individual thread running the code on lines 3–5 in Figure 2.2.

A compute kernel typically consists of thousands of threads, each of which starts by running the same function. In our example the CPU starts computation on the GPU on line 17 using CUDA's kernel configuration syntax. The kernel configuration syntax looks much like a function call in C with some additional information specifying the number of threads contained

```
1   void saxpy_serial(int n, float a, float *x, float *y)
2   {
3     for (int i = 0; i < n; ++i)
4       y[i] = a*x[i] + y[i];
5   }
6   main() {
7     float *x, *y;
8     int n;
9     // omitted: allocate CPU memory for x and y and initialize contents
10    saxpy_serial(n, 2.0, x, y); // Invoke serial SAXPY kernel
11    // omitted: use y on CPU, free memory pointed to by x and y
12  }
```

Figure 2.1: Traditional CPU code (based on Harris [2012]).

```
1   __global__ void saxpy(int n, float a, float *x, float *y)
2   {
3       int i = blockIdx.x*blockDim.x + threadIdx.x;
4       if(i<n)
5         y[i] = a*x[i] + y[i];
6   }
7   int main() {
8     float *h_x, *h_y;
9     int n;
10    // omitted: allocate CPU memory for h_x and h_y and initialize contents
11    float *d_x, *d_y;
12    int nblocks = (n + 255) / 256;
13    cudaMalloc( &d_x, n * sizeof(float) );
14    cudaMalloc( &d_y, n * sizeof(float) );
15    cudaMemcpy( d_x, h_x, n * sizeof(float), cudaMemcpyHostToDevice );
16    cudaMemcpy( d_y, h_y, n * sizeof(float), cudaMemcpyHostToDevice );
17    saxpy<<<nblocks, 256>>>(n, 2.0, d_x, d_y);
18    cudaMemcpy( h_x, d_x, n * sizeof(float), cudaMemcpyDeviceToHost );
19    // omitted: use h_y on CPU, free memory pointed to by h_x, h_y, d_x, and d_y
20  }
```

Figure 2.2: CUDA code (based on Harris [2012]).

between triple angle brackets (<<<>>>). The threads that make up a compute kernel are organized into a hierarchy composed of a *grid* of *thread blocks* consisting of *warps*. In the CUDA programming model, individual threads execute instructions whose operands are scalar values (e.g., 32-bit floating-point). To improve efficiency typical GPU hardware executes groups of threads together in lock-step. These groups are called *warps* by NVIDIA and *wavefronts* by AMD. NVIDIA warps consists of 32 threads while AMD wavefronts consist of 64 threads. Warps are grouped into a larger unit called a cooperative thread array (CTA) or thread block by NVIDIA. Line 17 indicates the compute kernel should launch a single grid consisting of `nblocks` thread blocks where each thread block contains 256 threads. The arguments passed by the CPU code to the kernel configuration statement are distributed to each instance of a running thread on the GPU.

Many of today's mobile device system-on-chips integrate a CPU and a GPU into a single chip as do processors found on today's laptop and desktop computers. However, traditionally, GPUs had their own DRAM memory and this continues today for GPUs found inside data centers used for machine learning. We note that NVIDIA has introduced Unified Memory, which transparently updates GPU memory from CPU memory and CPU memory from GPU memory. In systems enabled with Unified Memory the runtime and hardware are responsible for performing copies on the programmer's behalf. Given the increasing interest in machine learning and as the goal in this book is to understand hardware, in our example we consider the general case of separate GPU and CPU memories managed by the programmer.

Following the style used in many NVIDIA CUDA examples, we use the prefix `h_` in naming pointer variables for memory allocated in CPU memory and `d_` for pointers for memory allocated in GPU memory. On line 13 the CPU calls the CUDA library function `cudaMalloc`. This function invokes the GPU driver and asks it to allocate memory on the GPU for use by the program. The call to `cudaMalloc` sets `d_x` to point to a region of GPU memory containing enough space to hold `n` 32-bit floating-point values. On line 15 the CPU calls the CUDA library function `cudaMemcpy`. This function invokes the GPU driver and asks it to copy the contents of the array in CPU memory pointed to by `h_x` to the array in GPU memory pointed to by `d_x`.

Let us finally focus on the execution of threads on the GPU. A common strategy employed in parallel programming is to assign each thread a portion of the data. To facilitate this strategy, each thread on the GPU can lookup its own identity within the grid of blocks of threads. The mechanism for doing this in CUDA employs grid, block and thread identifiers. In CUDA, grids and thread blocks have x, y, and z dimensions. While it executes, each thread has a fixed, unique combination of non-negative integer x, y, and z coordinates within the grid and thread block. Each thread block has x, y, and z coordinates within a grid. Similarly, each thread has x, y, and z coordinates within a thread block. The extents of these coordinates are set by the kernel configuration syntax (line 17). In our example, y and z dimensions are not specified and so all threads have zero values for their y and z thread block and thread coordinates. On line 3 the value of `threadIdx.x` identifies the x coordinate of the thread within its thread block and `blockIdx.x` indicates the x coordinate of the thread block within its grid. The value `blockDim.x` indicates the maximum number of threads in the x-dimension. In our example, `blockDim.x` would evaluate to 256 since this is the value specified on line 17. The expression `blockIdx.x*blockDim.x + threadIdx.x` is used to compute an offset `i` for use when accessing the arrays `x` and `y`. As we will see, using index `i` we have assigned each thread a unique element of `x` and `y`.

To a significant extent, the combination of compiler and hardware enables the programmer to remain oblivious to the lock-step nature of thread execution in a warp. The compiler and hardware enable the appearance of each thread within a warp executing independently. On line 4 in Figure 2.2 we compare the value of index `i` with `n`, the size of the arrays `x` and `y`. Threads for which `i` is less than `n` execute line 5. Line 5 in Figure 2.2 performs one iteration of the original loop in Figure 2.1. After all threads in the grid are completed, the compute kernel returns control to the CPU after line 17. On line 18 the CPU invokes the GPU driver to copy the array pointed to by `d_y` from GPU memory back to CPU memory.

Some additional details of the CUDA programming model that are not illustrated by the SAXPY example, but which we will discuss later, are as follows.

Threads within a CTA can communicate with each other efficiently via a per compute core scratchpad memory. This scrathpad is called *shared memory* by NVIDIA. Each streaming multiprocessor (SM) contains a single shared memory. The space in the shared memory is divided up among all CTAs running on that SM. AMD's Graphics Core Next (GCN) architecture [AMD, 2012] includes a similar scratchpad memory that AMD calls the *local data store* (LDS). These scratchpad memories are small, ranging from 16–64 KB per SM, and exposed to programmers as different memory spaces. Programmers allocate memory into scratchpad memory using special keywords in their source code (e.g., "`__shared__`" in CUDA). The scratchpad memory acts as a software controlled cache. While GPUs also contain hardware managed caches, accessing data through such caches can lead to frequent cache misses. Applications benefit from using scratchpad memory when the programmer can identify data that is reused frequently and in a predictable manner. Unlike GPUs from NVIDIA, AMD's GCN GPUs also includes a *global data store* (GDS) scratchpad memory shared by all cores on the GPU. Scratchpad memories are used in graphics applications to pass results between different graphics shaders. For example, LDS is used for passing of parameter values betwen vertex and pixel shaders in GCN [AMD, 2012].

Threads within a CTA can synchronize efficiently using hardware-supported barrier instructions. Threads in different CTAs can communicate but do so through a global address space that is accessible to all threads. Access to this global address space is typically more expensive than access to shared memory both in terms of time and energy.

NVIDIA introduced CUDA Dynamic Parallelism (CDP) with the Kepler generation of GPUs [NVIDIA Corporation, a]. CDP is motivated by the observation that data intensive irregular applications can result in load imbalance across threads running on a GPU, leading the

GPU hardware to be underutilized. In many ways, the motivation is similar to that of Dynamic Warp Formation (DWF) [Fung et al., 2007] and related approaches discussed in Section 3.4.

2.2   GPU INSTRUCTION SET ARCHITECTURES

In this section, we briefly discuss the translation of compute kernels from a high-level language such as CUDA and OpenCL to the assembly level executed by the GPU hardware and the form of current GPU instruction sets. An interesting aspect of GPU architectures that is somewhat different from CPU architectures is the way the GPU ecosystem has evolved to support instruction set evolution. For example, x86 microprocessors are backwards compatible to the Intel 8086 released in 1976. Backward compatibility means that a program compiled for a prior generation architecture will run on the next generation architecture without any changes. Thus, software compiled 40 years ago for an Intel 8086 can in theory run on any of today's x86 processors.

2.2.1   NVIDIA GPU INSTRUCTION SET ARCHITECTURES

Given the, at times, large number of vendors offering GPU hardware (each with their own hardware designs), a level of instruction set virtualization, via OpenGL Shading Language (OGSL) and Microsoft's High-Level Shading Language (HLSL), became common as early GPUs became programmable. When NVIDIA introduced CUDA in early 2007, they decided to follow a similar path and introduced their own high-level virtual instruction set architecture for GPU computing called the Parallel Thread Execution ISA, or PTX [NVI, 2017]. NVIDIA fully documents this virtual instruction set architecture with each release of CUDA to the point that it was easy for the authors of this book to develop the GPGPU-Sim simulator to support PTX [Bakhoda et al., 2009]. PTX is many ways similar to a standard reduced instruction set computer (RISC) instruction set architecture like ARM, MIPS, SPARC, or ALPHA. It also shares a similarity to the intermediate representations used within optimizing compilers. One such example is the use of a limitless set of virtual registers. Figure 2.3 illustrates a PTX version of the SAXPY program from Figure 2.2.

Before running PTX code on the GPU it is necessary to compile PTX down to the actual instruction set architecture supported by the hardware. NVIDIA calls this level SASS which is short for "Streaming ASSembler" [Cabral, 2016]. The process of converting from PTX to SASS can be accomplished either by the GPU driver or a stand-alone program called ptxas provided with NVIDIA's CUDA Toolkit. NVIDIA does not fully document SASS. While this makes it harder for academic researchers to develop architecture simulators that capture all compiler optimization effects, it frees NVIDIA from customer demands to provide backwards compatibility at the hardware level enabling complete redesign of the instruction set architecture from one generation to the next. Inevitably, developers who wished to understand performance at a low level started to create their own tools to disassemble SASS. The first such effort due to Wladimir Jasper van der Laan and named "decuda" [van der Lann], arrived in late 2007 for NVIDIA's GeForce 8 Series (G80), within about a year of the first release of CUDA-enabled hardware.

```
1   .visible .entry _Z5saxpyifPfS_(
2   .param .u32 _Z5saxpyifPfS__param_0,
3   .param .f32 _Z5saxpyifPfS__param_1,
4   .param .u64 _Z5saxpyifPfS__param_2,
5   .param .u64 _Z5saxpyifPfS__param_3
6   )
7   {
8   .reg .pred %p<2>;
9   .reg .f32 %f<5>;
10  .reg .b32 %r<6>;
11  .reg .b64 %rd<8>;
12
13
14  ld.param.u32 %r2, [_Z5saxpyifPfS__param_0];
15  ld.param.f32 %f1, [_Z5saxpyifPfS__param_1];
16  ld.param.u64 %rd1, [_Z5saxpyifPfS__param_2];
17  ld.param.u64 %rd2, [_Z5saxpyifPfS__param_3];
18  mov.u32 %r3, %ctaid.x;
19  mov.u32 %r4, %ntid.x;
20  mov.u32 %r5, %tid.x;
21  mad.lo.s32 %r1, %r4, %r3, %r5;
22  setp.ge.s32 %p1, %r1, %r2;
23  @%p1 bra BB0_2;
24
25  cvta.to.global.u64 %rd3, %rd2;
26  cvta.to.global.u64 %rd4, %rd1;
27  mul.wide.s32 %rd5, %r1, 4;
28  add.s64 %rd6, %rd4, %rd5;
29  ld.global.f32 %f2, [%rd6];
30  add.s64 %rd7, %rd3, %rd5;
31  ld.global.f32 %f3, [%rd7];
32  fma.rn.f32 %f4, %f2, %f1, %f3;
33  st.global.f32 [%rd7], %f4;
34
35  BB0_2:
36  ret;
37  }
```

Figure 2.3: **PTX** code corresponding to compute kernel in Figure 2.2 (compiled with CUDA 8.0).

The decuda project developed a sufficiently detailed understanding of the SASS instruction set that it was possible to develop an assembler. This helped in developing support for SASS up to NVIDIA's GT200 architecture in GPGPU-Sim 3.2.2 [Tor M. Aamodt et al.]. NVIDIA eventually introduced a tool, called cuobjdump, and started to partially document SASS. NVIDIA's SASS documentation [NVIDIA Corporation, c] currently (April 2018) provides only a list of the assembly opcode names but no details on operand formats or SASS instruction semantics. More recently, with the explosive growth in the use of GPUs for machine learning and the need for performance-optimized code, others have developed tools similar to decuda for subsequent architectures such as NVIDIA's Fermi [Yunqing] and NVIDIA's Maxwell architecture [Gray].

Figure 2.4 illustrates SASS code for our SAXPY kernel compiled for NVIDIA's Fermi architecture [NVI, 2009] and extracted with NVIDIA's cuobjdump (part of the CUDA Toolkit). The first column in Figure 2.4 is the address of the instruction. The second column is assembly and the third column is the encoded instruction. As noted above, NVIDIA only partially documents their hardware assembly. Comparing Figure 2.3 and Figure 2.4, one can note both similarities and differences between the virtual and hardware ISA levels. At a high level there are important similarities such as both being RISC (both used loads and stores to access memory) and both using predication [Allen et al., 1983]. More subtle differences include: (1) the PTX version has an essentially infinite set of registers available so that each definition typically uses a new register much like static single assignment [Cytron et al., 1991] whereas SASS uses a limited set of registers; and (2) the kernel parameters are passed via banked constant memory which can be accessed by non load/store instructions in SASS, whereas parameters are allocated into their own separate "parameter" address space in PTX.

Figure 2.5 illustrates SASS code for SAXPY generated by the same version of CUDA but for NVIDIA's Pascal architecture and extracted with NVIDIA's cuobjdump. Comparing Figure 2.5 with Figure 2.4, it is clear NVIDIA's ISA has changed significantly including in terms of instruction encodings. Figure 2.5 contains some lines for which there is no disassembled instructions (e.g., at address 0x0000 on Line 3). These are special "control instructions" introduced in the NVIDIA Kepler architecture to eliminate the need for explicit dependency checking using a scoreboard [NVIDIA Corporation, b]. Lai and Seznec [2013] explored the encoding of control instructions for the Kepler architecture. As noted by Lai and Seznec [2013], these control instructions appear to be similar to the explicit-dependence lookahead on the Tera Computer System [Alverson et al., 1990]. Gray describes extensive details of the control instruction encoding that they were able to infer for NVIDIA's Maxwell architecture. According to Gray there is one control instruction for every three regular instructions in Maxwell. This appears to also be the case for NVIDIA's Pascal architecture as illustrated in Figure 2.5. According to Gray the 64-bit control instructions on Maxwell contain three groups of 21-bits encoding the following information for each of the following three instructions: a stall count; yeild hint flag; and write, read, and wait dependency barriers. Gray also describes the use of register reuse flags on regular instructions which can also be seen in Figure 2.5 (e.g., R0.reuse used for the first source

```
1  | Address        Dissassembly                                             Encoded Instruction
2  | ========       ==============================================           =========================
3  | /*0000*/           MOV R1, c[0x1][0x100];                               /* 0x2800440400005de4 */
4  | /*0008*/           S2R R0, SR_CTAID.X;                                  /* 0x2c00000094001c04 */
5  | /*0010*/           S2R R2, SR_TID.X;                                    /* 0x2c00000084009c04 */
6  | /*0018*/           IMAD R0, R0, c[0x0][0x8], R2;                        /* 0x2004400020001ca3 */
7  | /*0020*/           ISETP.GE.AND P0, PT, R0, c[0x0][0x20], PT;           /* 0x1b0e40008001dc23 */
8  | /*0028*/       @P0 BRA.U 0x78;                                          /* 0x40000001200081e7 */
9  | /*0030*/      @!P0 MOV32I R5, 0x4;                                      /* 0x18000000100161e2 */
10 | /*0038*/      @!P0 IMAD R2.CC, R0, R5, c[0x0][0x28];                    /* 0x200b8000a000a0a3 */
11 | /*0040*/      @!P0 IMAD.HI.X R3, R0, R5, c[0x0][0x2c];                  /* 0x208a8000b000e0e3 */
12 | /*0048*/      @!P0 IMAD R4.CC, R0, R5, c[0x0][0x30];                    /* 0x200b8000c00120a3 */
13 | /*0050*/      @!P0 LD.E R2, [R2];                                       /* 0x840000000020a085 */
14 | /*0058*/      @!P0 IMAD.HI.X R5, R0, R5, c[0x0][0x34];                  /* 0x208a8000d00160e3 */
15 | /*0060*/      @!P0 LD.E R0, [R4];                                       /* 0x8400000000402085 */
16 | /*0068*/      @!P0 FFMA R0, R2, c[0x0][0x24], R0;                       /* 0x3000400090202000 */
17 | /*0070*/      @!P0 ST.E [R4], R0;                                       /* 0x9400000000402085 */
18 | /*0078*/           EXIT;                                                /* 0x8000000000001de7 */
```

Figure 2.4: Low-level SASS code corresponding to compute kernel in Figure 2.2 (compiled with CUDA 8.0 for the NVIDIA Fermi Architecture, sm_20).

operand in the Integer Short Multiply Add instruction, XMAD, on Line 7). This appears to indicate an "operand reuse cache" was added in NVIDIA GPUs starting with Maxwell (see related research in Section 3.6.1). This operand reuse cache appears to enable register values to be read multiple times for each main register file access resulting in reduced energy consumption and/or improved performance.

2.2.2 AMD GRAPHICS CORE NEXT INSTRUCTION SET ARCHITECTURE

In contrast to NVIDIA when AMD introduced their Southern Islands architecture, they released a complete hardware-level ISA specification [AMD, 2012]. Southern Islands was the first generation of AMD's Graphics Core Next (GCN) architecture. The availability of documentation for AMD's hardware ISA has helped academic researchers in developing simulators that work at a lower level [Ubal et al., 2012]. AMD's compilation flow also includes a virtual instruction set architecture, called HSAIL, as part of the Heterogeneous System Architecture (HSA).

A key difference between AMD's GCN architecture and NVIDIA GPUs (including NVIDIA's most recent Volta architecture [NVIDIA Corp., 2017]) is separate scalar and vector instructions. Figures 2.6 and 2.7 reproduce an example from AMD [2012] of high-level OpenCL (similar to CUDA) code and the equivalent machine instructions for the AMD South-

```
Address        Dissassembly                                          Encoded Instruction
========       ==============================================        =========================
                                                                     /* 0x001c7c00e22007f6 */
/*0008*/       MOV R1, c[0x0][0x20];                                 /* 0x4c98078000870001 */
/*0010*/       S2R R0, SR_CTAID.X;                                   /* 0xf0c8000002570000 */
/*0018*/       S2R R2, SR_TID.X;                                     /* 0xf0c8000002170002 */
                                                                     /* 0x001fd840fec20ff1 */
/*0028*/       XMAD.MRG R3, R0.reuse, c[0x0] [0x8].H1, RZ;           /* 0x4f107f8000270003 */
/*0030*/       XMAD R2, R0.reuse, c[0x0] [0x8], R2;                  /* 0x4e00010000270002 */
/*0038*/       XMAD.PSL.CBCC R0, R0.H1, R3.H1, R2;                   /* 0x5b30011800370000 */
                                                                     /* 0x081fc400ffa007ed */
/*0048*/       ISETP.GE.AND P0, PT, R0, c[0x0][0x140], PT;           /* 0x4b6d038005070007 */
/*0050*/   @P0 EXIT;                                                 /* 0xe30000000000000f */
/*0058*/       SHL R2, R0.reuse, 0x2;                                /* 0x3848000000270002 */
                                                                     /* 0x081fc440fec007f5 */
/*0068*/       SHR R0, R0, 0x1e;                                     /* 0x3829000001e70000 */
/*0070*/       IADD R4.CC, R2.reuse, c[0x0][0x148];                  /* 0x4c10800005270204 */
/*0078*/       IADD.X R5, R0.reuse, c[0x0][0x14c];                   /* 0x4c10080005370005 */
                                                                     /* 0x0001c800fe0007f6 */
/*0088*/       IADD R2.CC, R2, c[0x0][0x150];                        /* 0x4c10800005470202 */
/*0090*/       IADD.X R3, R0, c[0x0][0x154];                         /* 0x4c10080005570003 */
/*0098*/       LDG.E R0, [R4];            }                          /* 0xeed4200000070400 */
                                                                     /* 0x0007c408fc400172 */
/*00a8*/       LDG.E R6, [R2];                                       /* 0xeed4200000070206 */
/*00b0*/       FFMA R0, R0, c[0x0][0x144], R6;                       /* 0x4980030005170000 */
/*00b8*/       STG.E [R2], R0;                                       /* 0xeedc200000070200 */
                                                                     /* 0x001f8000ffe007ff */
/*00c8*/       EXIT;                                                 /* 0xe30000000007000f */
/*00d0*/       BRA 0xd0;                                             /* 0xe2400fffff87000f */
/*00d8*/       NOP;                                                  /* 0x50b0000000070f00 */
                                                                     /* 0x001f8000fc0007e0 */
/*00e8*/       NOP;                                                  /* 0x50b0000000070f00 */
/*00f0*/       NOP;                                                  /* 0x50b0000000070f00 */
/*00f8*/       NOP;                                                  /* 0x50b0000000070f00 */
```

Figure 2.5: Low-level SASS code corresponding to compute kernel in Figure 2.2 (compiled with CUDA 8.0 for the NVIDIA Pascal Architecture, sm_60).

ern Islands architecture. In Figure 2.7, scalar instructions are prefaced with s_ and vector instructions are prefaced with v_. In the AMD GCN architecture, each compute unit (e.g., SIMT core) contains a scalar unit coupled with four vector units. Vector instructions execute on the vector units and compute different 32-bit values for each individual thread in a wavefront. In contrast, scalar instructions execute on the scalar units compute a single 32-bit value shared by all threads in a wavefront. In the example shown in Figure 2.7 the scalar instructions are related to control flow handling. In particular, exec is a special register used to predicate execution of individual vector lanes for SIMT execution. The use of masking for control flow handling on GPUs is described in more detail in Section 3.1.1. Another potential benefit of the scalar unit in the GCN architecture is that frequently certain portions of a computation in a SIMT program will compute the same result independent of thread ID (see Section 3.5).

```
1   float fn0(float a,float b)
2   {
3       if(a>b)
4           return(a * a - b);
5       else
6           return(b * b - a);
7   }
```

Figure 2.6: OpenCL code (based on Figure 2.2 in AMD [2012]).

```
1    // Registers r0 contains "a", r1 contains "b"
2    // Value is returned in r2
3        v_cmp_gt_f32 r0, r1 // a>b
4        s_mov_b64 s0, exec  // Save current exec mask
5        s_and_b64 exec, vcc, exec // Do "if"
6        s_cbranch_vccz label0 // Branch if all lanes fail
7        v_mul_f32 r2, r0, r0 // result = a * a
8        v_sub_f32 r2, r2, r1 // result = result - b
9    label0:
10       s_not_b64 exec, exec // Do "else"
11       s_and_b64 exec, s0, exec // Do "else"
12       s_cbranch_execz label1 // Branch if all lanes fail
13       v_mul_f32 r2, r1, r1 // result = b * b
14       v_sub_f32 r2, r2, r0 // result = result - a
15   label1:
16       s_mov_b64 exec, s0   // Restore exec mask
```

Figure 2.7: Southern Islands (graphics core next) microcode (based on Figure 2.2 in AMD [2012]).

AMD's GCN hardware instruction set manual [AMD, 2012] provides many interesting insights into AMD GPU hardware. For example, to enable data dependency resolution for long latency operations AMD's GCN architecture includes s_WAITCNT instructions. For each wavefront there are three counters: vector memory count, local/global data store count, and register export count. Each of these indicate the number of outstanding operations of a given type. The compiler or programmer inserts s_WAITCNT instructions to have the wavefront wait until the number of outstanding operations decreases below a specified threshold.

CHAPTER 3

The SIMT Core: Instruction and Register Data Flow

In this and the following chapter we will examine the architecture and microarchitecture of modern GPUs. We divide our discussion of GPU architecture into two parts: (1) examining the SIMT cores that implement computation in this chapter and then (2) looking at the memory system in the next chapter.

In their traditional graphics-rendering role, GPUs access data sets such as detailed texture maps that are far too large to be fully cached on-chip. To enable high-performance programmability, which is desirable in graphics both to ease verification costs as the number of graphics modes increase and to enable games developers to more easily differentiate their products [Lindholm et al., 2001], it is necessary to employ an architecture that can sustain large off-chip bandwidths. Thus, today's GPUs execute tens of thousands of threads concurrently. While the amount of on-chip memory storage per thread is small, caches can still be effective in reducing a sizable number of off-chip memory accesses. For example, in graphics workloads, there is significant spatial locality between adjacent pixel operations that can be captured by on-chip caches.

Figure 3.1 illustrates the microarchitecture of the GPU pipeline discussed in this chapter. This figure illustrates the internal organization of a single SIMT-core shown in Figure 1.2. The pipeline can be divided into a SIMT front-end and a SIMD back-end. The pipeline consists of three scheduling "loops" acting together in a single pipeline: an instruction fetch loop, an instruction issue loop, and a register access scheduling loop. The instruction fetch loop includes the blocks labeled Fetch, I-Cache, Decode, and I-Buffer. The instruction issue loop includes the blocks labeled I-Buffer, Scoreboard, Issue, and SIMT Stack. The register access scheduling loop includes the blocks labeled Operand Collector, ALU, and Memory. In the rest of this chapter we help you build up a full understanding of the individual blocks in this picture by considering key aspects of the architecture that depend on each of these loops.

As there are many details involved in fully understanding this organization, we divide our discussion up in parts. We order these with the objective of developing an increasingly detailed view of the core microarchitecture. We start with a high-level view of the overall GPU pipeline and then fill in details. We call these increasingly accurate descriptions "approximations" to acknowledge that some details are omitted even in our most detailed descriptions. As the central organizing principle of today's GPUs is multithreading we organize these "approximations"

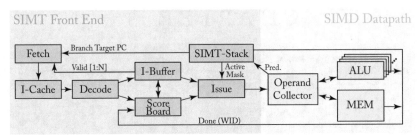

Figure 3.1: Microarchitecture of a generic GPGPU core.

around the three scheduling loops described above. We have found it convenient to organize this chapter by considering three increasingly accurate "approximation loops" that progressively take into account the details of these scheduler loops.

3.1 ONE-LOOP APPROXIMATION

We start by considering a GPU with a single scheduler. This simplified look at the hardware is not unlike what one might expect the hardware to do if they only read the description of the hardware found in the CUDA programming manual.

To increase efficiency, threads are organized into groups called "warps" by NVIDIA and "wavefronts" by AMD. Thus, the unit of scheduling is a warp. In each cycle, the hardware selects a warp for scheduling. In the one loop approximation the warp's program counter is used to access an instruction memory to find the next instruction to execute for the warp. After fetching an instruction, the instruction is decoded and source operand registers are fetched from the register file. In parallel with fetching source operands from the register file, the SIMT execution mask values are determined. The following sub-section describes how the SIMT execution mask values are determined and contrasts them with predication, which is also employed in modern GPUs.

After the execution masks and source registers are available, execution proceeds in a single-instruction, multiple-data manner. Each thread executes on the function unit associated with a lane provided the SIMT execution mask is set. As in modern CPU designs, the function units are typically heterogeneous meaning a given function unit supports only a subset of instructions. For example, NVIDIA GPUs contain a *special function unit* (SFU), *load/store unit, floating-point function unit, integer function unit*, and, as of Volta, a *Tensor Core*.

All function units nominally contain as many lanes as there are threads within a warp. However, several GPUs have used a different implementation in which a single warp or wavefront is executed over several clock cycles. This is achieved by clocking the function units at a higher frequency, which can achieve higher performance per unit area at the expense of increased energy consumption. One way to achieve higher clock frequencies for the function units is to pipeline their execution or increase their pipeline depth.

3.1.1 SIMT EXECUTION MASKING

A key feature of modern GPUs is the SIMT execution model, which from the standpoint of functionality (although not performance) presents the programmer with the abstraction that individual threads execute completely independently. This programming model can potentially be achieved via predication alone. However, in current GPUs it is achieved via a combination of traditional predication along with a stack of predicate masks that we shall refer to as the *SIMT stack*.

The SIMT stack helps efficiently handle two key issues that occur when all threads can execute independently. The first is nested control flow. In nested control flow one branch is control dependent upon another. The second issue is skipping computation entirely while all threads in a warp avoid a control flow path. For complex control flow this can represent a significant savings. Traditionally, CPUs supporting predication have handled nested control flow by using multiple predicate registers and supporting across lane predicate tests has been proposed in the literature.

The SIMT stack employed by GPUs can handle both nested control flow and skipped computation. There are several implementations described in patents and instruction set manuals. In these descriptions the SIMT stack is at least partly managed by special instructions dedicated to this purpose. Instead, we will describe a slightly simplified version introduced in an academic work that assumes the hardware is responsible for managing the SIMT stack.

To describe the SIMT stack we use an example. Figure 3.2 illustrates CUDA C code that contains two branches nested within a do-while loop and Figure 3.3 illustrates the corresponding PTX assembly. Figure 3.4, which reproduces Figure 5 in Fung et al. [Fung et al., 2007], illustrates how this code interacts with the SIMT stack assuming a GPU that has four threads per warp.

Figure 3.4a illustrates a control flow graph (CFG) corresponding to the code in Figures 3.2 and 3.3. As indicated by the label "A/1111" inside the top node of the CFG, initially all four threads in the warp are executing the code in Basic Block A which corresponds to the code on lines 2 through 6 in Figure 3.2 and lines 1 through 6 in Figure 3.3. These four threads follow different (divergent) control flow after executing the branch on line 6 in Figure 3.3, which corresponds to the "if" statement on line 6 in Figure 3.2. Specifically, as indicated by the label "B/1110" in Figure 3.4a the first three threads fall through to Basic Block B. These three threads branch to line 7 in Figure 3.3 (line 7 in Figure 3.2). As indicated by the label "F/0001" in Figure 3.4a, after executing the branch the fourth thread jumps to Basic Block F, which corresponds to line 14 in Figure 3.3 (line 14 in Figure 3.2).

Similarly, when the three threads executing in Basic Block B reach the branch on line 9 in Figure 3.3 the first thread diverges to Basic Block C while the second and third thread diverges to Basic Block D. Then, all three threads reach Basic Block E and execute together as indicated by the label "E/1110" in Figure 3.4a. At Basic Block G all four threads execute together.

How does GPU hardware enable threads within a warp to follow different paths through the code while employing a SIMD datapath that allows only one instruction to execute per cycle?

```
1    do {
2        t1 = tid*N;        // A
3        t2 = t1 + i;
4        t3 = data1[t2];
5        t4 = 0;
6        if( t3 != t4 ) {
7          t5 = data2[t2]; // B
8          if( t5 != t4 ) {
9            x += 1;        // C
10         } else {
11           y += 2;        // D
12         }
13       } else {
14         z += 3;          // F
15       }
16       i++;               // G
17   } while( i < N );
```

Figure 3.2: Example CUDA C source code for illustrating SIMT stack operation.

```
1    A:   mul.lo.u32      t1, tid, N;
2         add.u32         t2, t1, i;
3         ld.global.u32   t3, [t2];
4         mov.u32         t4, 0;
5         setp.eq.u32     p1, t3, t4;
6    @p1  bra             F;
7    B:   ld.global.u32   t5, [t2];
8         setp.eq.u32     p2, t5, t4;
9    @p2  bra             D;
10   C:   add.u32         x, x, 1;
11        bra             E;
12   D:   add.u32         y, y, 2;
13   E:   bra             G;
14   F:   add.u32         z, z, 3;
15   G:   add.u32         i, i, 1;
16        setp.le.u32     p3, i, N;
17   @p3  bra             A;
```

Figure 3.3: Example PTX assembly code for illustrating SIMT stack operation.

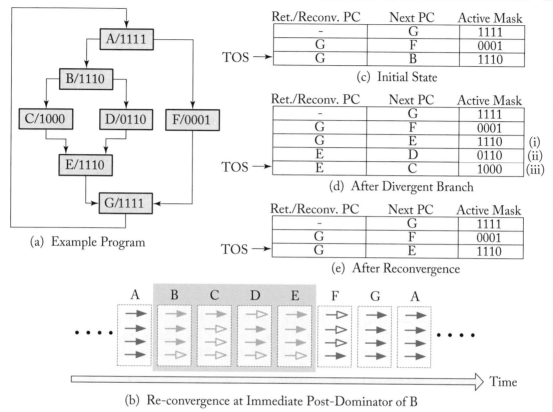

(a) Example Program

Ret./Reconv. PC	Next PC	Active Mask
-	G	1111
G	F	0001
G	B	1110

TOS → (third row)

(c) Initial State

Ret./Reconv. PC	Next PC	Active Mask	
-	G	1111	
G	F	0001	
G	E	1110	(i)
E	D	0110	(ii)
E	C	1000	(iii)

TOS → (fifth row)

(d) After Divergent Branch

Ret./Reconv. PC	Next PC	Active Mask
-	G	1111
G	F	0001
G	E	1110

TOS → (third row)

(e) After Reconvergence

(b) Re-convergence at Immediate Post-Dominator of B

Figure 3.4: Example of SIMT stack operation (based on Figure 5 from Fung et al. [2007]).

The approach used in current GPUs is to serialize execution of threads following different paths within a given warp. This is illustrated in Figure 3.4b where arrows represent threads. A filled-in arrow indicates the thread is executing the code in the corresponding basic block (indicated by the letter on top each rectangle). An arrow with a hollow head indicates the thread is masked off. Time advances to the right in the diagram, as indicated by the arrow at the bottom. Initially, each thread is executing in Basic Block B. Then, after the branch the first three threads execute the code in Basic Block B. Note that at this point in time thread four is masked off. To maintain SIMD execution the fourth thread executes the alternate code path through Basic Block F at a different time (several cycles later in this example).

To achieve this serialization of divergent code paths one approach is to use a stack like that is illustrated in Figure 3.4c–e. Each entry on this stack contains three entries: a reconvergence program counter (RPC), the address of the next instruction to execute (Next PC), and an active mask.

Figure 3.4c illustrates the state of the stack immediately after the warp has executed the branch on line 6 in Figure 3.3. Since three threads branch to Basic Block B and one thread branches to Basic Block F, two new entries have been added to the top of the stack (TOS). The next instruction that the warp executes is determined using the Next PC value in the top of stack (TOS) entry. In Figure 3.4c, this Next PC value is B, which represents the address for the first instruction in Basic Block B. The corresponding Active Mask entry, "1110", indicates only the first three threads in the warp should execute this instruction.

The first three threads in the warp continue executing instructions from Basic Block B until they reach the branch on line 9 in Figure 3.3. After executing this branch they diverge, as noted earlier. This branch divergence causes three changes to the stack. First, the Next PC entry of the TOS entry prior to executing the branch, labeled (i) in Figure 3.4d, is modified to the *reconvergence point* of the branch, which is the address of the first instruction in Basic Block E. Then, two entries, labeled (ii) and (iii) in Figure 3.4d, are added, one for each of the paths followed by threads in the warp after executing the branch.

A reconvergence point is a location in the program where threads that diverge can be forced to continue executing in lock-step. The nearest reconvergence point is generally preferred. The earliest point in a given program execution where it can be guaranteed at compile time that threads which diverge can again execute in lock-step is the immediate post-dominator of the branch that caused the branch divergence. At runtime it is sometimes possible to reconverge at an earlier point in the program [Coon and Lindholm, 2008, Diamos et al., 2011, Fung and Aamodt, 2011].

An interesting question is "what order should be used to add the entries to the stack following a divergent branch?" To reduce the maximum depth of the reconvergence stack to be logarithmic in the number of threads in a warp it is best to put the entry with the most active threads on the stack first and then the entry with fewer active threads [AMD, 2012]. In part (d) of Figure 3.4 we follow this order while in part (c) we used the opposite order.

3.1.2   SIMT DEADLOCK AND STACKLESS SIMT ARCHITECTURES

Recently, NVIDIA has disclosed details of their upcoming Volta GPU architecture [NVIDIA Corp., 2017]. One change they have highlighted will be to the behavior of masking under divergence and how this interacts with synchronization. The stack-based implementation of SIMT can lead to a deadlock condition called "SIMT deadlock" by ElTantawy and Aamodt [2016]. Academic work has described alternative hardware for SIMT execution [ElTantaway et al., 2014] that, with minor changes [ElTantawy and Aamodt, 2016], can avoid SIMT deadlock. NVIDIA calls their new thread divergence management approach Independent Thread Scheduling. The descriptions of independent thread scheduling suggest they achieve behavior similar to that obtained by the above academic proposals. Below, we first describe the SIMT deadlock problem, then we describe a mechanism for avoiding SIMT deadlock that is consistent with NVIDIA's description of independent thread scheduling and that was disclosed in a recent NVIDIA patent application [Diamos et al., 2015].

The left part of Figure 3.5 gives a CUDA example illustrating the SIMT deadlock problem and the middle part shows the corresponding control flow graph. Line A initializes the shared variable, mutex, to zero to indicate that a lock is free. On line B, each thread in a warp executes the atomicCAS operation, which performs a compare-and-swap operation on the memory location containing mutex. The atomicCAS operation is a compiler intrinsic that is translated to a atom.global.cas PTX instruction. Logically, the compare-and-swap first reads the contents of mutex, then it compares it to the second input, 0. If the current value of mutex is 0, then the compare and swap operation updates the value of mutex to the third input, 1. The value returned by atomicCAS is original value of mutex. Importantly, the compare-and-swap performs the above sequence of logical operations atomically for each thread. Thus, multiple accesses by atomicCAS to any single location, made by different threads within the same warp, are serialized. As all threads in Figure 3.5 access the same memory location, only one thread will see the value of mutex as 0, and the remaining threads will see the value as 1. Next, while keeping the SIMT-stack in mind, consider what happens with the while loop on line B after atomicCAS returns. Different threads see different loop conditions. Specifically, one thread will want to exit the loop while the remaining threads will want to stay in the loop. The thread that exits the loop will have reached the reconvergence point and thus will no longer be active on the SIMT-stack and thus unable to execute the atomicExch operation to release the lock on line C. The threads that remain in the loop will be active at the top of the SIMT-stack and will spin indefinitely. The resulting circular dependence between threads introduces a new form of deadlock, called SIMT-deadlock by ElTantawy and Aamodt [2016] that would not exist had the threads executed on a MIMD architecture.

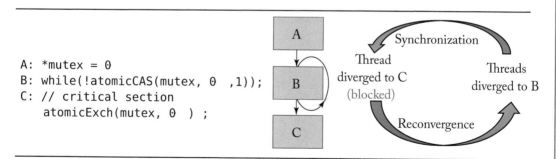

```
A: *mutex = 0
B: while(!atomicCAS(mutex, 0 ,1));
C: // critical section
   atomicExch(mutex, 0 ) ;
```

Figure 3.5: SIMT deadlock example (based on Figure 1 from ElTantawy and Aamodt [2016]).

Next, we summarize a stackless branch reconvergence mechanism like that in a recent US Patent Application from NVIDIA [Diamos et al., 2015]. This mechanism is consistent with NVIDIA's descriptions to date of Volta's reconvergence handling mechanisms [Nvidia, 2017]. The key idea is to replace the stack with per warp convergence barriers. Figure 3.6 shows various fields maintained per warp as described in NVIDIA's patent application and Figure 3.8

provides a corresponding example to illustrate the operation of convergence barriers. Effectively, the proposal provides an alternative implementation of Multi-Path IPDOM [ElTantaway et al., 2014], which will be described in Section 3.4.2 along with earlier academic works. The convergence barrier mechanism has some similarities to the notion of a *warp barrier* described in Fung and Aamodt [2011]. To help explain the convergence barrier mechanism below we consider the execution of a single warp on the code in Figure 3.8, which shows the control flow graph that results from CUDA code like that shown in Figure 3.7.

Barrier Participation Mask 425		
Barrier State 430		

Thread State 440-0	. . .	Thread State 440-31
Thread rPC 445-0	. . .	Thread rPC 445-31
Thread Active 460-0	. . .	Thread Active 460-31

Figure 3.6: Alternate stack-less convergence barrier based branch divergence handling mechanism recently described by NVIDIA (based on Figure 4B from Diamos et al. [2015]).

```
1   // id = warp ID
2   // BB_A  Basic Block "A"
3   if(id%2==0){
4       // BB_B
5   }else{
6       // BB_C
7       if(id==1){
8           // BB_D
9       }else{
10          // BB_E
11      }
12      // BB_F
13  }
14  // BB_G
```

Figure 3.7: Nested control flow example (based on Figure 6(a) from ElTantaway et al. [2014]).

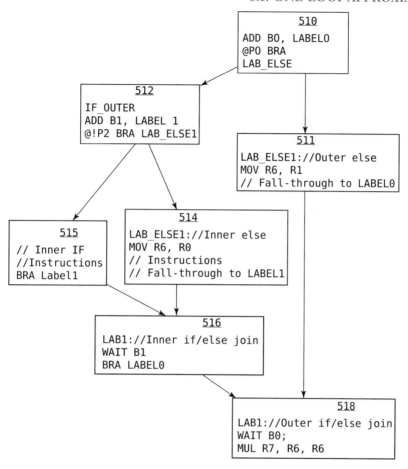

Figure 3.8: Code example for convergence barrier branch divergence handling mechanism recently described by NVIDIA (based on Figure 5B from Diamos et al. [2015]).

Next, we describe the fields in Figure 3.6. These fields are stored in registers and used by the hardware warp scheduler. Each *Barrier Participation Mask* is used to track which threads within a given warp participate in a given convergence barrier. There may be more than one barrier participation mask for a given warp. In the common case threads tracked by a given barrier participation mask will wait for each other to reach a common point in the program following a divergent branch and thereby reconverge together. To support this the *Barrier State* field is used to track which threads have arrived at a given convergence barrier. The *Thread State* tracks, for each thread in the warp whether the thread is ready to execute, blocked at a convergence barrier (and if so, which one), or has yielded. It appears the yielded state is may be used to enable other threads in the warp to make forward progress past the convergence barrier in a situation that would otherwise lead to SIMT deadlock. The Thread rPC field tracks, for each thread that is not active, the address of the next instruction to execute. The Thread Active field is a bit that indicates if the corresponding thread in the warp is active.

Assuming a warp contains 32 threads, the barrier participation mask is 32-bits wide. If a bit is set, that means the corresponding thread in the warp participates in this convergence barrier. Threads diverge when they execute a branch instruction such as those at the end of basic blocks 510 and 512 in Figure 3.8. These branches correspond to the two "if" statements in Figure 3.7. The barrier participation mask is used by the warp scheduler to stop threads at a specific convergence barrier location which can be the immediate postdominator of the branch or another location. At any given time each warp may require multiple barrier participation masks to support nested control flow constructs such as the nested if statements in Figure 3.7. The registers in Figure 3.6 might be implemented using general-purpose registers or dedicated registers or some combination of the two (the patent application does not say). Given the barrier participation mask is only 32-bits wide, it would be redundant if each thread had a copy of the barrier participation mask as it might if naively using the general-purpose register file to store it. However, as the control flow can be nested to arbitrary depths, a given warp may need an arbitrary number of barrier participation masks making software management of the mask desirable.

To initialize the convergence barrier participation mask a special "ADD" instruction is employed. All threads that are active when the warp executes this ADD instruction have their bit set in the convergence barrier indicated by the ADD instruction. After executing a branch some threads may diverge, meaning the address of the next instruction (i.e., PC) to execute will differ. When this happens the scheduler will select a subset of threads with a common PC and update the Thread Active field to enable execution for these threads of the warp. Academic proposals refer to such a subset of threads as a "warp split" [ElTantaway et al., 2014, ElTantawy and Aamodt, 2016, Meng et al., 2010]. In contrast to a stack-based SIMT implementation, with a convergence barrier implementation the scheduler is free to switch between groups of diverged threads. This enables forward progress between threads in a warp when some threads have acquired a lock while others have not.

A "WAIT" instruction is used to stop a warp split when it reaches a convergence barrier. As described in NVIDIA's patent application, the WAIT instruction includes an operand to indicate the identity of the convergence barrier. The effect of the WAIT instruction is to add the threads in the warp split to the Barrier State register for the barrier and change the threads' state to blocked. Once all threads in the barrier participation mask have executed the corresponding WAIT instruction the thread scheduler can switch all the threads from the original warp split to active and SIMD efficiency is maintained. The example in Figure 3.8 has two convergence barriers, B1 and B2 with WAIT instructions in basic blocks 516 and 518. To enable switching between warp splits NVIDIA describes using a YIELD instruction along with other details such as support for indirect branches that we omit in this discussion [Diamos et al., 2015].

Figure 3.9 shows an example of the timing of stack-based reconvergence and Figure 3.10 illustrates potential timing using independent thread scheduling as described in NVIDIA's Volta whitepaper. In Figure 3.10, we can see statements A and B are interleaved with statements X and Y by Volta in contrast with the behavior in Figure 3.9. This behavior is consistent with the convergence barrier mechanism described above (as well as Multi-Path IPDOM [ElTantaway et al., 2014]). Finally, Figure 3.11 illustrates how a stackless architecture might execute the spin look code from Figure 3.5 so as to avoid SIMT deadlock.

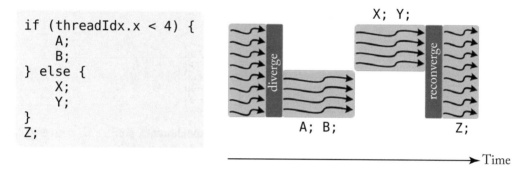

Figure 3.9: Example showing behavior of stack-based reconvergence (based on Figure 20 from Nvidia [2017]).

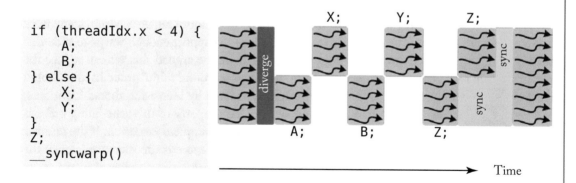

Figure 3.10: Example showing behavior of Volta reconvergence (based on Figure 23 from Nvidia [2017]).

3.1.3   WARP SCHEDULING

Each core in a GPU hosts contains many warps. A very interesting question is which order these warps should be scheduled in. To simplify the discussion we assume that each warp issues only a single instruction when it is scheduled and furthermore that the warp is not eligible

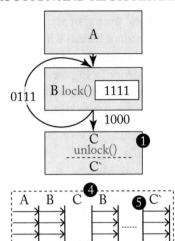

Figure 3.11: Example showing behavior of an academic mechanism similar to convergence-barrier on spin lock code from Figure 3.5 (based on Figure 6(a) from ElTantawy and Aamodt [2016]).

to issue another instruction until the first instruction completes execution. We will revisit this assumption later in this chapter.

If the memory system were "ideal" and responded to memory requests within some fixed latency it would, in theory, be possible to design the core to support enough warps to hide this latency using fine-grained multithreading. In this case it can be argued that we can reduce the area of the chip for a given throughput by scheduling warps in "round robin" order. In round robin the warps are given some fixed ordering, for example ordered by increasing thread identifiers, and warps are selected by the scheduler in this order. One property of this scheduling order is that it allows roughly equal time to each issued instruction to complete execution. If the number of warps in a core multiplied by the issue time of each warp exceeds the memory latency the execution units in the core will always remain busy. So, increasing the number of warps up to this point can in principle increase throughput per core.

However, there is an important trade off: to enable a different warp to issue an instruction each cycle it is necessary that each thread have its own registers (this avoids the need to copy and restore register state between registers and memory). Thus, increasing the number of warps per core increases the fraction of chip area devoted to register file storage relative to the fraction dedicated to execution units. For a fixed chip area increasing warps per core will decrease the total number of cores per chip.

In practice, the response latency of memory depends upon the application's locality properties and the amount of resulting contention encountered by off-chip memory accesses. What impact does scheduling play when considering the memory system of the GPU? This has been the topic of considerable research in the past few years and we will return to this question after adding more detail about the memory system to our model of GPU microarchitecture. However, briefly, locality properties can either favor or discourage round-robin scheduling: when different threads share data at a similar point in their execution, such as when accessing texture maps in graphics pixel shaders, it is beneficial for threads to make equal progress as this can increase the number of memory references which "hit" in on-chip caches, which is encouraged by round-robin scheduling [Lindholm et al., 2015]. Similarly, accessing DRAM is more efficient when nearby locations in the address space are accessed nearby in time and this is also encouraged by round-robin scheduling [Narasiman et al., 2011]. On the other hand, when threads mostly access disjoint data, as tends to occur with more complex data structures, it can be beneficial for a given thread to be scheduled repeatedly so as to maximize locality [Rogers et al., 2012].

3.2   TWO-LOOP APPROXIMATION

To help reduce the number of warps that each core must support to hide long execution latencies it is helpful to be able to issue a subsequent instruction from a warp while earlier instructions have not yet completed. However, the one-loop microarchitecture described earlier prevents this because the scheduling logic in that design only has access to the thread identifier and the address of the next instruction to issue. Specifically, it does not know whether the next instruction to issue for the warp has a dependency upon an earlier instruction that has not yet completed execution. To provide such dependency information it is necessary to first fetch the instruction from memory so as to determine what data and/or structural hazards exists. For this purpose, GPUs implement an instruction buffer were instructions are placed after cache access. A separate scheduler is used to decide which of several instructions in the instruction buffer should be issued next to the rest of the pipeline.

Instruction memory is implemented as a first-level instruction cache backed by one or more levels of secondary (typically unified) caches. The instruction buffer can also help in hiding instruction cache miss latencies in combination with instruction miss-status holding registers (MSHRs) [Kroft, 1981]. After a cache hit or a fill from a cache miss, the instruction information is placed into the instruction buffer. The organization of the instruction buffer can take many forms. One particularly straightforward approach is to have storage for one or more instructions per warp.

Next, let us consider how to detect data dependencies between instructions within the same warp. Their are two traditional approaches to detecting dependencies between instructions found in traditional CPU architectures: a scoreboard and reservation stations. Reservation stations are used for eliminating name dependencies and introduce the need for associative logic that is expensive in terms of area and energy. Scoreboards can be designed to support either in-order execution or out-of-order execution. Scoreboards supporting out-of-order execution, like that used in the CDC 6600, are also fairly complex. On the other hand, the scoreboard for a single threaded in-order CPU is very simple: each register is represented in the scoreboard with a single bit that is set whenever an instruction issues that will write to that register. Any instruction that wants to read or write to a register that has its corresponding bit set in the scoreboard is stalled until the bit is cleared by the instruction writing to the register. This prevents both read-after-write and write-after-write hazards. When combined with in-order instruction issue this simple scoreboard can prevent write-after-read hazards provided that reading of the register file is constrained to occur in-order which is typically the case in in-order CPU designs. Given it is the simplest design and therefore will consume the least amount of area and energy, GPUs implement in-order scoreboards. However, as discussed next, there are challenges to using an in-order scoreboard when supporting multiple warps.

The first concern with the simple in-order scoreboard design described above is the very large number of registers contained in modern GPUs. With up to 128 registers per warp and up to 64 warps per core a total of 8192 bits per core is required to implement the scoreboard.

Another concern with the simple in-order scoreboard design described above is that an instruction that encounters a dependency must repeatedly lookup its operands in the scoreboard until the prior instruction it depends upon writes its results to the register file. With a single-threaded design this introduces little complexity. However, in an in-order issue multithreaded processor instructions from multiple threads may be waiting for earlier instructions to complete. If all such instructions must probe the scoreboard additional read ports are required. Recent GPUs support up to 64 warps per core and with up to 4 operands allowing all warps to probe the scoreboard every cycle would require 256 read ports, which would be very expensive. One alternative would be to restrict the number of warps that can probe the scoreboard each cycle, but this restricts the number of warps that can be considered for scheduling. Also, if none of the instructions checked are free of dependencies no instruction may be issued even if other instructions that could not be checked happened to be free of dependencies.

Both issues can be addressed using a design proposed by Coon et al. [2008]. Rather than hold a single bit per register per warp, the design contains a small number (estimated to be around 3 or 4 in one recent study [Lashgar et al., 2016]) of entries per warp, where each entry is the identifier of a register that will be written by an instruction that has been issued but not yet completed execution. A regular in-order scoreboard is accessed both when instructions issue and when they write back. Instead, Coon et al.'s scoreboard is accessed when an instruction is placed into the instruction buffer and when an instruction writes its results into the register file.

When an instruction is fetched from the instruction cache and placed in the instruction buffer the scoreboard entries for the corresponding warp are compared against that instructions' source and destination registers. This results in a short bit vector, with one bit for each entry in the scoreboard for that warp (e.g., 3 or 4 bits). A bit is set if the corresponding entry in the scoreboard matched any of the operands of the instruction. This bit vector is then copied alongside the instruction in the instruction buffer. An instruction is not eligible to be considered by the instruction scheduler until all bits are cleared, which can be determined by feeding each bit of the vector into a NOR gate. Dependency bits in the instruction buffer are cleared as instructions write their results to the register file. If all entries are used up for a given warp then either fetch stalls for all warps or the instruction is discarded and must be fetched again. When an instruction that has executed is ready to write to the register file it clears the entry that was allocated to it in the scoreboard and also clears the corresponding dependency bit for any instructions from the same warp that are stored in the instruction buffer.

In the two-loop architecture, the first loop selects a warp that has space in the instruction buffer, looks up its program counter and performs an instruction cache access to obtain the next instruction. The second loop selects an instruction in the instruction buffer that has no outstanding dependencies and issues it to the execution units.

3.3   THREE-LOOP APPROXIMATION

As described earlier, to hide long memory latencies it is necessary to support many warps per core and to support cycle by cycle switching between warps it is necessary to have a large register file that contains separate physical registers for every warp that is executing. For example, such register contain 256 KB on recent GPU architectures from NVIDIA (e.g., Kepler, Maxwell, and Pascal architectures). Now, the area of an SRAM memory is proportional to the number of ports. A naive implementation of a register file requires one port per operand per instruction issued per cycle. One way to reduce the area of the register file is to simulate the large number of ports using multiple banks of single-ported memories. While it is possible to achieve such effects by exposing these banks to the instruction set architecture, in some GPU designs it appears a structure known as the operand collector [Coon et al., 2009, Lindholm et al., 2008b, Lui et al., 2008] is used to achieve this in a more transparent way. The operand collector effective forms a third scheduling loop as described below.

To better understand the problem solved by the operand collector, first consider Figure 3.12, which shows a naive microarchitecture for providing increased register file bandwidth. This figure shows the register read stage of a GPU instruction pipeline where the register file is composed for four single-ported logical banks of registers. In practice, as the register file is very large, each logical bank may be further decomposed into a larger number of physical banks (not shown). The logical banks are connected via a crossbar to staging registers (labeled "pipeline register") that buffer source operands before passing them to a SIMD execution unit. An arbiter controls access to the individual banks and routes results through a crossbar to the appropriate staging register.

Figure 3.13 shows a naive layout of each warp's registers to logical banks. In this figure, register r0 from warp 0 (w0) is stored in the first location in Bank 0, register r1 from warp 0 is stored in the first location in Bank 1, and so on. If the number of registers required by the computation is larger than the number of logical banks, the allocation wraps around. For example, register r4 for warp 0 is stored in the second location in Bank 0.

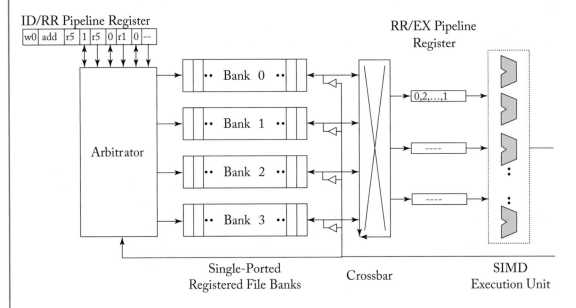

Figure 3.12: Naive banked register file microarchitecture.

Bank 0	Bank 1	Bank 2	Bank 3
...	...	...	...
w1:r4	w1:r5	w1:r6	w1:r7
w1:r0	w1:r1	w1:r2	w1:r3
w0:r4	w0:r5	w0:r6	w0:r7
w0:r0	w0:r1	w0:r2	w0:r3

Figure 3.13: Naive banked register layout.

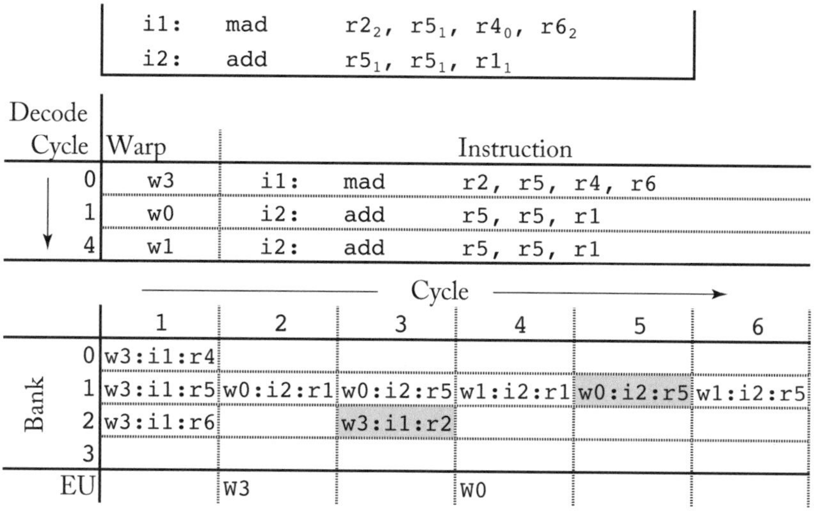

```
i1:    mad      r2₂, r5₁, r4₀, r6₂
i2:    add      r5₁, r5₁, r1₁
```

Decode Cycle	Warp	Instruction			
0	w3	i1:	mad	r2, r5, r4, r6	
1	w0	i2:	add	r5, r5, r1	
4	w1	i2:	add	r5, r5, r1	

		← Cycle →					
		1	2	3	4	5	6
Bank	0	w3:i1:r4					
	1	w3:i1:r5	w0:i2:r1	w0:i2:r5	w1:i2:r1	w0:i2:r5	w1:i2:r5
	2	w3:i1:r6		w3:i1:r2			
	3						
EU			w3		w0		

Figure 3.14: Timing of naive banked register file.

Figure 3.14 illustrates a timing example that highlights how this microarchitecture cause hurt performance. The example involves two instructions shown at the top. The first instruction, i1, is a multiple-add operation that reads from registers r5, r4, and r6 which are allocated in banks 1, 0, and 2 (indicated by subscripts in the figure). The second instruction, i2, is an addition instruction that reads from registers r5 and r1 both allocated in bank 1. The middle part of the figure shows the order that instructions are issued. On cycle 0 warp 3 issues instruction i1, on cycle 1 warp 0 issues instruction i2 and on cycle 4 warp 1 issues instruction i2 after a delay due to bank conflicts as described next. The bottom portion of the figure illustrates the timing of accesses to different banks by the different instructions. On cycle 1, instruction i1 from warp 3 is able to read all three of its source registers on cycle 1 because they map to distinct logical banks. However, on cycle 2, instruction i2 from warp 0 is only able to read one of its two source registers because both map to bank 1. On cycle 3, the second source register for this instruction is read in parallel with the writeback of instruction i1 from warp 3. On cycle 4, instruction i2 from warp 1 is able to read its first-source operand but not the second as, again, both map to bank 1. On cycle 5, the second-source operand from instruction i2 from warp 1 is prevented from being read from the register file due to the fact the bank is already being accessed by the higher-priority writeback of instruction i2 issued earlier by warp 0. Finally, on cycle 6 the second source operand of i2 from warp 1 is read from the register file. In summary, it takes six cycles for three instructions to finish reading their source registers and during this time many of the banks are not accessed.

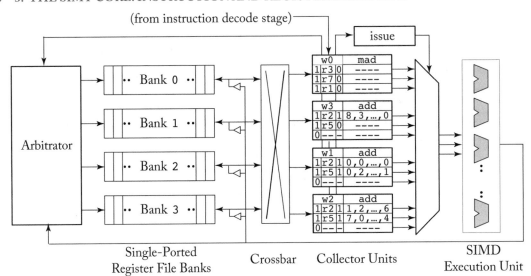

Figure 3.15: Operand collector microarchitecture (based on Figure 6 from Tor M. Aamodt et al.).

3.3.1   OPERAND COLLECTOR

The operand collector microarchitecture [Lindholm et al., 2008b] is illustrated in Figure 3.15. The key change is that the staging registers have been replaced with *collector units*. Each instruction is allocated a collector unit when it enters the register read stage. There are multiple collector units so that multiple instructions can overlap reading of source operands which can help improve throughput in the presence of bank conflicts between the source operands of individual instructions. Each collector unit contains buffering space for all source operands required to execute an instruction. Given the larger number of source operands for multiple instructions the arbiter is more likely to achieve increased bank-level parallelism to allow accessing multiple register file banks in parallel.

The operand collector uses scheduling to tolerate bank conflicts when they occur. This leaves open the question of how to reduce the number of bank conflicts. Figure 3.16 illustrates a modified register layout that Coon et al. describe for helping to reduce bank conflicts. The idea is to allocate equivalent registers from different warps in different banks. For example, in Figure 3.16 register r0 for warp 0 is allocated to bank 0, but register r0 for warp 1 is allocated to bank 1. This does not address bank conflicts between register operands of a single instruction. However, where it does help is in reducing bank conflicts between instructions from different warps. In particular, whenever warps are making relatively even progress (e.g., due to round-robin scheduling or two-level scheduling [Narasiman et al., 2011] in which individual warps in a fetch group are scheduled in round-robin order).

Bank 0	Bank 1	Bank 2	Bank 3
...	...	...	...
w1:r7	w1:r4	w1:r5	w1:r6
w1:r3	w1:r0	w1:r1	w1:r2
w0:r4	w0:r5	w0:r6	w0:r7
w0:r0	w0:r1	w0:r2	w0:r3

Figure 3.16: Swizzled banked register layout.

```
i1:  add   r1, r2, r5
i2:  mad   r4, r3, r7, r1
```

Cycle	Warp	Instruction
0	w1	i1: add $r1_2$, $r2_3$, $r5_2$
1	w2	i1: add $r1_3$, $r2_0$, $r5_3$
2	w3	i1: add $r1_0$, $r2_1$, $r5_0$
3	w0	i2: mad $r4_0$, $r3_3$, $r7_3$, $r1_1$

Cycle →

Bank	1	2	3	4	5	6
0		w2:r2		w3:r5		w3:r1
1			w3:r2			
2		w1:r5		w1:r1		
3	w1:r2		w2:r5	w0:r3	w2:r1	w0:r7
EU			w1	w2	w3	

Figure 3.17: Timing of operand collector.

Figure 3.17 shows a timing example with a sequence of addition and multiply-add instructions shown at the top. In the middle the issue order is shown. Three instances of i1 from warps 1 through 3 are issued on cycles 0 through 2. An instance of instruction i2 from warp 0 issues on cycle 3. Notice that the add instructions write to register r1, which for any given warp is allocated in the same bank as source register r5. However, unlike the case using the register layout in Figure 3.13, here different warps access different banks which helps reduce conflicts between writeback of one warp and reading source operands in other warps. The bottom portion shows the bank level timing of accesses due to the operand collector. On cycle 1 register r2 from warp 1 reads Bank 3. On cycle 4, notice the writeback of register r1 from warp 1 proceeds in parallel with reading register r5 from warp 3 and reading register r3 from warp 0.

A subtle issue with the operand collector as described so far is that because it does not impose any order between when different instructions are ready to issue, it may allow write-after-read (WAR) hazards [Mishkin et al., 2016]. This can occur if two instructions from the same warp are present in an operand collector with the first instruction reading a register that the second instruction will write. If the first instruction's source operand accesses encounter repeated bank conflicts the second instruction can conceivably write a new value to the register before the first register has read the (correct) older value. One way to prevent this WAR hazard is simply to require that instructions from the same warp leave the operand collector to execution units in program order. Mishkin et al. [2016] explore three potential solutions with low hardware complexity and evaluate their performance impact. The first, a *release-on-commit warpboard*, allows at most one instruction per warp to be executing. Unsurprisingly, they find this impacts performance negatively, reducing performance by almost a factor of two in some cases. Their second proposal is a *release-on-read warpboard* which allows only one instruction at a time per warp to be collecting operands in the operand collector. This scheme results in slowdowns of at most 10% on the workloads they studied. Finally, to allow for instruction level parallelism in the operand collector they propose a *bloomboard* mechanism that uses a small bloom filter to track outstanding register reads. This results in impact of less than a few percent vs. (incorrectly) allowing WAR hazards. Separately, an analysis performed by Gray suggests NVIDIA's Maxwell GPU introduced a "read dependency barrier" which is managed by special "control instructions" and which may be used to avoid WAR hazards for certain instructions (see Section 2.2.1).

3.3.2 INSTRUCTION REPLAY: HANDLING STRUCTURAL HAZARDS

There are many potential causes of structural hazards in GPU pipelines. For example, the register read stage may run out of operand collector units. Many sources of structural hazards relate to the memory system, which we will discuss in more detail the next chapter. In general, a single memory instruction executed by a warp may need to be broken down into multiple separate operations. Each of these separate operations may fully utilize a portion of the pipeline on a given cycle.

What happens when an instruction encounters a structural hazard in a GPU pipeline? In a single-threaded in-order CPU pipeline a standard solution is to stall younger instructions until the instruction encountering the stall condition can make further progress. This approach is arguable less desirable in a highly multithreaded throughput architecture for at least two reasons. First, given the large size of the register file along with the many pipeline stages required to support a full graphics pipeline distributing a stall signal may impact the critical path. Pipelining stall-cycle distribution leads to the need to introduce additional buffering increasing area. Second, stalling an instruction from one warp may cause instructions from other warps to stall behind it. If those instructions do not require the resource required by the instruction that caused the stall, throughput may suffer.

To avoid these issues GPUs implement a form of instruction replay. Instruction replay is found in some CPU designs where it is used as a recovery mechanism when speculatively scheduling a dependent instruction upon a earlier instruction that has variable latency. For example, loads may either hit or miss in a first-level cache but CPU designs that are clocked at a high frequency may pipeline first-level cache access over as many as four clock cycles. Some CPUs speculative wake up instructions depending upon a load so as to improve single threaded performance. In contrast, GPUs avoid speculation as it tends to waste energy and reduce throughput. Instead, instruction replay is used in GPUs to avoid clogging the pipeline and the circuit area and/or timing overheads resulting from stalling.

To implement instruction replay a GPU can hold instructions in the instruction buffer either until it is known that they have completed or all individual portions of the instruction have executed [Lindholm et al., 2015].

3.4   RESEARCH DIRECTIONS ON BRANCH DIVERGENCE

This section is based on Wilson Fung's Ph.D. dissertation [Fung, 2015].
Ideally, threads within the same warp execute through the same control flow path, so that the GPU can execute them in lockstep on SIMD hardware. Given the autonomy of the threads, a warp may encounter a *branch divergence* when its threads diverge to different targets at a data-dependent branch. Modern GPUs contains special hardware to handle branch divergence in a warp. Section 3.1.1 describes the baseline SIMT stack, which is used by the baseline GPU architecture in this book. The baseline SIMT stack handles branch divergence in a warp by serializing the execution of the different targets. While the baseline SIMT stack handles the branch divergence correctly for most existing GPU applications, it has the following deficiencies.

Lower SIMD Efficiency In the presence of branch divergence, the baseline SIMT stack serializes the execution of each branch target. As each target is executed, the SIMT stack only activates the subset of scalar threads running the target. This causes some lanes in the SIMD hardware to be idle, lowering the overall *SIMD efficiency*.

Needless Serialization The serialized execution of each branch target by the baseline SIMT stack is not required for functional correctness. The GPU programming model does not impose any implicit data dependency between scalar threads within a warp—they have to communicate explicitly via Shared Memory and barrier. The GPU can interleave the execution of all branch targets of a diverged warp to make use of idle cycles in the SIMD hardware.

Inadequate MIMD Abstraction By forcing diverged warps to reconverge at a compiler-defined reconvergence point, the baseline SIMT stack implicitly imposes a warp-wide synchronization point at each reconvergence point. This works for many existing GPU applications. However, this implicit synchronization may interact pathologically with other user-implemented synchronization mechanisms, such as fine-grained locks, causing the warp to deadlock. The compiler-defined reconvergence point also does not account for the control-flow divergence introduced by system-level constructs such as exceptions and interrupts.

Area Cost While the area requirement of the baseline SIMT stack for each warp is only 32 ×
64 bits (or as low as 6 × 64 bits), the area scales with the number of in-flight warps in
the GPU. In typical GPU applications where branch divergence is rare, the SIMT stack
occupies area that can otherwise be used to boos the application throughput in other ways
(e.g., large cache, more ALU units, etc.).

Both the industry and academia has proposed alternatives to address the above deficiencies. The various proposals can be classified into the following categories: warp compaction,
intra-warp divergent path management, adding MIMD capability and complexity reduction.
Some proposals contain improvements that capture aspects from multiple categories, and thus
are mentioned multiple times.

3.4.1 WARP COMPACTION

With GPU implementing fine-grained multi-threading to tolerate long-memory access latency,
there are many warps in each SIMT core, hundreds to thousands of scalar threads in total. Since
these warps are usually running the same compute kernel, they are likely to follow the same execution path, and encounter branch divergence at the same set of data-dependent branches. Con-
sequently, each target of a divergent branch is probably executed by a large number of threads,
but these threads are scattered among multiple static warps, with each warp handling the divergence individually.

In this section, we summarize a series of research that exploits this observation to improve
the performance of GPU applications that suffer from branch divergence. Proposals in this series
all involve novel hardware mechanisms to *compact* threads from different *static warps* into new
dynamic warps to improve the overall SIMD efficient of these divergent GPU applications. Here,
static warps refers to the warps that are formed by the GPU hardware when the scalar threads
are spawned from a kernel launch. In our baseline GPU architecture, this arrangement is fixed
throughout the execution of the warp. The arrangement of scalar threads into static warps is an
arbitrary grouping imposed by the GPU hardware that is largely invisible to the programming
model.

Dynamic Warp Formation.   Dynamic warp formation (DWF) [Fung et al., 2007, Fung et al.,
2009] exploits this observation by rearranging these scattered threads that execute the same instruction into new dynamic warps. At a divergent branch, DWF can boost the overall SIMD efficiency of an application by compacting threads scattered among multiple diverged static warps
into fewer non-divergent dynamic warps. In this way, DWF can capture a significant fraction
of the benefits of MIMD hardware on SIMD hardware. However, DWF requires warps to encounter the same divergent branch within a short time window. This timing-dependent nature
of DWF makes it very sensitive to the warp scheduling policy.

The follow-up work by Fung and Aamodt [2011] identifies two major of performance
pathologies for DWF: (1) a greedy scheduling policy can starve some threads, leading to a SIMD

efficiency reduction; and (2) Thread regrouping in DWF increases non-coalesced memory accesses and shared memory bank conflicts. These pathologies cause DWF to slowdown many existing GPU applications. Moreover, applications that rely on implicit synchronization in a static warp execute incorrectly with DWF.

The above pathologies can be partially addressed with an improved scheduling policy that effectively separates the compute kernel into two sets of regions: divergent and non-divergent (coherent) regions. The divergent regions benefit significantly from DWF, whereas the coherent regions are free of branch divergence but are prone to the DWF pathologies. We found that the impact of the DWF pathologies can be significantly reduced by forcing DWF to rearrange scalar threads back to their static warps in the coherent regions.

Thread Block Compaction.   Thread block compaction (TBC) [Fung and Aamodt, 2011] builds upon this insight with the observation that rearrangement of threads into new dynamic warps continually does not yield additional benefit. Instead, the rearrangement, or *compaction*, only needs to happen right after a divergent branch, the start of a divergent region, and before its reconvergence point, the start to a coherent region. We note the existing per-warp SIMT stack (described in Chapter 3.1.1) implicitly synchronizes threads diverged to different execution paths at the reconvergence point of the divergent branch, merging these diverged threads back into a static warp before executing a coherent region. TBC extends the SIMT stack to encompass all warps executing in the same core, forcing them to synchronize and compact at divergent branches and reconvergence points to achieve robust DWF performance benefits. However, synchronizing all the warps within a core at each divergent branch for compaction can greatly reduce the available thread-level parallelism (TLP). GPU architectures rely on the abundance of TLP to tolerate pipeline and memory latency.

TBC settles for a compromise between SIMD efficiency and TLP availability by restricting compaction to only occur within a *thread block*. GPU applications usually execute multiple thread blocks concurrently on a single core to overlap the synchronization and memory latency. TBC leverages this software optimization to overlap the compaction overhead at divergent branches—when warps in one thread block synchronize for compaction at a divergent branch, warps in other thread blocks can keep the hardware busy. It extends the per-warp SIMT stack to encompass warps in a thread block. The warp scheduling logic uses this thread-block-wide SIMT stack to determine when warps in a thread block should synchronizes and be compacted into a new sets of warps. The result is a far more robust and simple mechanism that captures much of the benefits of DWF, without the pathological behaviors.

Large Warp Microarchitecture.   The large warp microarchitecture [Narasiman et al., 2011] extends the SIMT stack, similar to TBC, to manage the reconvergence of a group of warps. However, instead of restricting the compaction at branches and reconvergence points. LWM requires warps within the group to execute in complete lockstep, so that it can compact the group at every instruction. This reduces the available TLP even more so than TBC, but allows

LWM to perform compaction with predicated instructions as well as unconditional jumps. Similar to TBC, LWM splits warps running on the same core into multiple groups, and restricts compaction to occur only within a group. It also opts for a more complex scoreboard microarchitecture that tracks register dependency at thread-granularity. This allows some warps in the group to execute slightly ahead of others compensate the lost TLP due to lockstep execution.

Compaction-Adequacy Predictor.   Rhu and Erez [2012] extend TBC with a compaction-adequacy predictor (CAPRI). The predictor identifies the effectiveness of compacting threads into few warps at each branch, and only synchronizes the threads at branches where the compaction is predicted to yield a benefit. This reclaims the TLP lost due to non-beneficial stall and compaction with TBC. Rhu and Erez [2012] also show that a simple history-based predictor similar to a single-level branch predictor is sufficient to achieve high accuracy.

Intra-Warp Compaction.   Vaidya et al. [2013] propose a low-complexity compaction technique that benefits wide SIMD execution groups that executes multiple cycle on narrower hardware units. Their basic technique divides a single execution group into multiple subgroups that match the hardware width. SIMD execution group that suffers from divergence can run faster on the narrow hardware by skipping subgroups that are completely idle. To create more completely idle subgroups, they propose a swizzle mechanism that compacts elements into fewer subgroups at divergence.

Simultaneous Warp Interweaving.   Brunie et al. [2012] propose simultaneous branch and warp interweaving (SBI and SWI). They extend the GPU SIMT front-end to support issuing two different instructions per cycle. They compensate this increased complexity by widening the warp to twice its original size. SWI co-issues an instruction from a warp suffering from divergence with instructions from another diverged warp to fill the gaps left by branch divergence.

Impact on Register File Microarchitecture

To avoid introducing extra communication traffics between SIMT cores, the hardware compaction proposals usually takes place locally within a SIMT core. Since the compacted threads all located on the same core sharing the same register file, it is possible to perform compaction without moving their architectural states with a more flexible register file design [Fung et al., 2007].

As discussed earlier in this chapter, GPU register files are implemented with large single-ported SRAM banks to maximize their area efficiency. Registers for threads in the same warp are stored in consecutive regions in the same SRAM bank, so that they can be accessed together via a single wide port. This allows high bandwidth register file access while amortizing register file access control hardware. Hardware warp compaction creates dynamic warps that may not obey this arrangement of registers. Fung et al. [2007] propose a more flexible register file design featuring SRAM banks with narrow ports. This design has more SRAM banks to maintain the same bandwidth.

Dynamic Micro-Kernels. Steffen and Zambreno [2010] improved SIMD efficiency of ray tracing on GPUs with *dynamic micro-kernels*. The programmer is given primitives to break iterations in a data-dependent loop into successive micro-kernel launches. This decomposition by itself does not improve parallelism, because each iteration depends on data from the previous iteration. Instead, the launch mechanism improves the load imbalance between different threads in the same core by compacting the remaining active threads into few warps. It also differs from the rest of the hardware warp compaction techniques in that the compaction migrates the threads with their architectural states, using the per-core scratchpad memory as a staging area.

Section 3.4.1 summarizes a series of research that implements warp compaction in software, which does not require the more flexible register file design. Instead, these proposals introduces extra memory traffic to relocate threads from one SIMT core to another.

Warp Compaction in Software

On existing GPUs, one way to improve SIMD efficiency of an application is through software warp compaction—using software to group threads/work items according to their control flow behavior. The regrouping involves moving the thread and its private data in memory, potentially introducing a significant memory bandwidth overhead. Below we highlight several works on software compaction techniques.

Conditional streams [Kapasi et al., 2000] apply this concept to stream computing. It splits a compute kernel for stream processors with potentially divergent control flow into multiple kernels. At a divergent branch, a kernel splits its data stream into multiple streams according to branch outcome of each data element. Each stream is then processed by a separate kernel, and merges back at the end of the control flow divergence.

Billeter et al. [2009] proposed to use a parallel prefix sum to implement SIMD *stream compaction*. The stream compaction reorganizes streams of elements with assorted tasks into compact substreams of identical tasks. This implementation leverages the access flexibility of the GPU on-chip scratchpad to achieve high efficiency. Hoberock et al. [2009] proposed a deferred shading technique for ray tracing that uses stream compaction to improve the SIMD efficiency of pixel shading in a complex scene with many material classes. Each material class requires its unique computation. A pixel shader combining the computation for every material class runs inefficiently on GPUs. Stream compaction groups the rays hitting objects with similar material classes, allowing the GPU SIMD hardware to execute the shader for these pixels efficiently.

Zhang et al. [2010] proposed a runtime system that remaps thread into different warps on the fly to improve SIMD efficiency as well as memory access spatial locality. The runtime system features a pipelined system, with the CPU performing the on-the-fly remapping and the GPU performing computations on the remapped data/threads.

Khorasani et al. [2015] proposed *Collective Context Collection* (CCC), a compiler technique that transforms a given GPU compute kernel with potential branch divergence penalty to improve its SIMD efficiency on existing GPUs. CCC focuses on compute kernels in which each thread performs an irregular amount of computation at each step, such as a breath-first-search through a irregular graph. Instead of assigning one node (or task in other applications) per thread, CCC first transforms the compute kernel so that each thread processes multiple nodes, with the node to warp (note: not thread) assignment determined ahead of the kernel launch. CCC then transforms the compute kernel so that each thread in a warp can offload the context of a task to a warp-specific stack stored in the shared memory. A warp that experiences low SIMD efficiency at its current set of tasks can offload the tasks to the stack, and uses these offloaded tasks to fill up threads that went idle they process a later set of tasks. In effect, CCC performs "warp compaction" by grouping the tasks from multiple warps into a fewer set of warps, and then compacting the divergent tasks into fewer iterations within each warp via the warp-specific stack stored in the fast, on-chip, shared memory.

Impacts of Thread Assignment within a Warp

In the baseline GPU architecture studied throughout this book, threads with consecutive thread IDs are statically fused together to form warps. Little academic work has gone into the static assignment of threads to warps or lanes in a warp. This default sequential mapping works well for most workloads since adjacent threads tend to access adjacent data, improving memory coalescing. However, some research has looked at alternatives.

SIMD Lane Permutation.   Rhu and Erez [2013b] make the observation that the sequential mapping of thread IDs to consecutive threads in a warp is suboptimal for warp compaction techniques described earlier in this section. A key limitation of most warp compaction and formation work is that when threads are assigned to a new warp, they cannot be assigned to a different lane, or else their register file state would have to be moved to a different lane in the vector register. Rhu and Erez observe that the structure of a program biases certain control flow paths to certain SIMD lanes. This biasing makes it more difficult to achieve compaction, since threads that take the same path tend to be in the same lane, preventing those threads from being merged together. Rhu and Erez propose several different thread mapping permutations that remove these programmatic biases and significantly improve the rate of compaction.

Intra-warp Cycle Compaction.   Vaidya et al. [2013] exploit the fact that the width of the SIMD datapath does not always equal the warp width. For example, in NVI [2009], the SIMD width is 16, but the warp size is 32. This means a 32-thread warp is executed over 2 core cycles. Vaidya et al. [2013] observe that when divergence occurs, if a sequential SIMD-worth of threads are masked off for an instruction, then the instruction can be issued in only one cycle, skipping the masked off lanes. They call their technique cycle compression. However, if the masked-off threads are not contiguous, the basic technique does not yield any performance improvement. To address this, they propose a swizzled cycle compression that re-arranges which threads are in which lanes in order to create more opportunities for cycle compression.

Warp Scalarization.   Other work, such as that by [Yang et al., 2014], argues that the SIMT programming model is inefficient when the threads with a warp operate on the same data. A number of solutions propose including a scalar unit in the pipeline for work that the compiler or programmer can identify as scalar a-priori. AMD's Graphics Core Next (GCN) architecture includes a scalar pipeline for this purpose. See Section 3.5 for more details.

3.4.2   INTRA-WARP DIVERGENT PATH MANAGEMENT

While a SIMT stack with immediate post-dominator reconvergence points can handle branch divergence with arbitrary control flow, it can be further improved in various aspects.

1. Threads diverged to different branch targets of a diverged warp can interleave their execution to make use of idle cycles in the SIMD hardware.

2. While the immediate post-dominator of a divergent branch is the definite convergence point, threads diverged to different branch targets may be able to converge before the immediate post-dominator of the divergent branch.

The following subsections highlight several works that attempt to improve the SIMT stack in these two aspects.

Multi-Path Parallelism

When a warp diverges at a branch, the threads are split into multiple groups, called *warp-splits*. Each warp-split consists of threads following the same branch target. In the baseline, *single path*, SIMT stack, warp-splits from the same warp are executed one-by-one, until the warp-split reaches its reconvergence point. This serialization lends itself to a relatively simple hardware implementation, but is not necessary for functional correctness. Threads in the warp have independent registers, and communicate among themselves explicitly via memory operations and synchronization operations such as barriers. In other words, every warp-split from the same warp may execute in *parallel*. We call this mode of execution *multi-path execution mode*.

While the different warp-splits may not execute on the same hardware at the same cycle (afterall, they runs different instructions), they may interleave their execution on the same hardware just like multiple warps interleaving their execution on the same datapath. In this way, multi-path execution mode boosts thread-level-parallelism (TLP) available in an application to tolerate memory access latency. Even though the SIMD efficiency is not improved, multi-path execution boosts the overall performance of memory-bound applications, where the SIMT cores has plenty of idle cycles to be filled with useful work.

Example 3.1 shows a short compute kernel that may benefit from multi-path execution. In this example, the code paths in both branch targets contain a load from memory. In the single-path SIMT stack, each of block B and C are executed serially until the corresponding warp-split reaches block D (the reconvergence point), even when the warp-split stalled waiting for data from memory. This stalls the entire warp, introducing idle cycles in the datapath, as shown in Figure 3.18, that has to be filled by works from other warps. With multi-path execution, the warp-splits of block B and C can interleave their execution, eliminating these idle cycles introduced by memory accesses.

Algorithm 3.1 Example of multi-path parallelism with branch divergence.

```
X = data[i];        // block A
if( X > 3 )
   result = Y[i] * i;  // block B
else
   result = Z[i] + i;  // block C
return result;      // block D
```

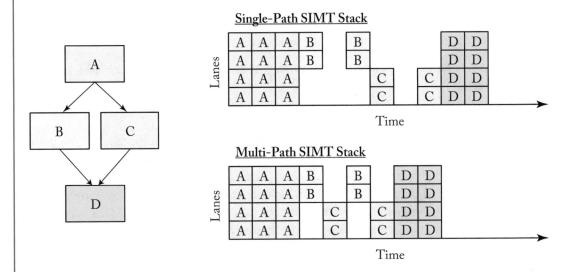

Figure 3.18: Multi-path execution at branch divergence.

Dynamic Warp Subdivision.   Meng et al. [2010] propose that *dynamic warp subdivision* (DWS) is the first proposal to exploit on the TLP boost from multi-path execution. DWS extends the SIMT stack with a warp-split table to subdivide a diverged warp into concurrent warp-splits. The warp-splits, each executing a divergent branch target, can execute in parallel to reclaim hardware idleness due to memory accesses. Warp-splits are also created at memory divergences—when only a portion of the threads in a warp hit in the L1 data cache. Instead of waiting for all threads to obtain their data, DWS split the warp and allow the warp-split that hits in the cache to execute ahead, potentially prefetching data for those who have missed the cache.

Dual-Path Execution.   Rhu and Erez [2013a] propose the dual-path SIMT stack (DPS), which addresses some of the implementation shortcomings of DWS by restricting each warp to execute only two concurrent warp-splits. While this restriction enables DPS to capture a good portion of the full DWS advantage, it leads to a far simpler hardware design. DPS only requires extending the baseline SIMT stack with an extra set of PC and active mask to encode the extra warp-split. Only the two warp-splits at the top-of-stack entry of a warp are executed in parallel; every other warp-split in the same warp is paused until its entry reaches top-of-stack. DPS is also accompanied with an extension of the scoreboard to track the register dependency of each warp-split independently. This allows a dual-path execution model to achieve greater TLP than DWS with the baseline scoreboard.

Multi-Path Execution.   ElTantaway et al. [2014] remove the dual-path limitation with a multi-path execution model (MPM). MPM replaces the SIMT stack with two tables: a warp-split table maintaining the set of warp-splits from the diverged warp, and a reconvergence table synchronizing all the warp-splits with the same reconvergence point.

At a divergent branch, one new entry is created in the reconvergence table, with the reconvergence point of the divergent branch (its immediate post-dominator). Multiple (usually two) entries is created in the warp-split table, one for each warp-split. Each warp-split entry maintains the current PC of the warp-split, its active mask, the reconvergence PC (RPC), and the R-index pointing to the corresponding entry in the reconvergent table. Every warp-split in the warp-split table is available for execution, until its PC == RPC. At this point, the corresponding reconvergence table entry is updated to reflect that threads from this warp-split has arrived at the reconvergence. When all pending threads have arrived at the reconvergence point, the reconvergence table entry is deallocated, and a new warp-split entry is created with the reconverged threads active, starting at the RPC.

MPM also extended the scoreboard to track the register dependency for each thread, without fully duplicating the scoreboard for each thread (which would render MPM impractical due to the significant area overhead in doing so). This is a crucial extension that allows the warp-splits to execute in a truly independent manner—without the extension, register dependency for one warp-split may be mistaken for the dependency of another warp-split from the same warp.

MPM is further extended with opportunistic early reconvergence, boosting the SIMD efficiency for unstructured control flow (see Section 3.4.2).

DWS, as well as other techniques discussed in this section, are orthogonal the warp compaction techniques discussed in Section 3.4.1. For example, the block-wide SIMT stack in TBC can be extended with DWS to boost the available TLP.

Better Convergence

The post-dominator (PDOM) stack-based reconvergence mechanism [Fung et al., 2007, Fung et al., 2009] uses reconvergence points identified using a unified algorithm rather than by translating control flow idioms in the source code into instructions [AMD, 2009, Coon and

Lindholm, 2008, Levinthal and Porter, 1984]. The immediate post-dominator of a divergent branch selected as the reconvergence point is the earliest point in a program where the divergent threads are *guaranteed* to reconverge. In certain situations, threads can reconverge at an *earlier point*, and if hardware can exploit this, it would improve SIMD efficiency. We believe this observation motivates the inclusion of the break instruction in recent NVIDIA GPUs [Coon and Lindholm, 2008].

The code in Example 3.2 (from [Fung and Aamodt, 2011]) exhibits this earlier reconvergence. It results in the control flow graph in Figure 3.19 where edges are marked with the probability with which individual scalar threads follow that path. Block F is the immediate post-dominator of A and C since F is the first location where *all* paths starting at A (or C) coincide. In the baseline mechanism, when a warp diverges at A, the reconvergence point is set to F. However, the path from C to D is rarely followed and hence in *most* cases threads can reconverge earlier at E.

Algorithm 3.2 Example for branch reconvergence earlier than immediate post-dominator.

```
while (i < K) {
  X = data[i];      // block A
  if( X == 0 )
    result[i] = Y;  // block B
  else if ( X == 1 ) // block C
    break;          // block D
  i++;              // block E
}
return result[i];   // block F
```

Likely-Convergence Points. Fung and Aamodt [2011] propose extending the SIMT stack with *likely convergence points*. This extension adds two new fields to each SIMT stack entry: one for the PC of the likely convergence point (LPC) and the other (LPos), a pointer that records the stack position of a special likely convergence entry created when a branch has a likely convergence point that differs from the immediate post-dominator. The likely convergence point of each branch can be identified with either control flow analysis or profile information (potentially collected at runtime). The proposal by Fung and Aamodt [2011] restrictes likely convergence points to the closest enclosing backward-taken branch to capture the impact of "break" statements within loops [Coon and Lindholm, 2008].

When a warp diverges at a branch with a likely-convergence point, three entries are pushed onto the SIMT stack. The first entry, an LPC entry, is created for the likely convergence point of the branch. Two other entries for the taken and fall through of the branch are created as in the baseline mechanism. The LPC field in each of these other entries is populated with the likely convergence point of the divergent branch, the LPos field populated with the stack position of the LPC entry. The LPC entry has its RPC set to the immediate post-dominator, i.e., the definite

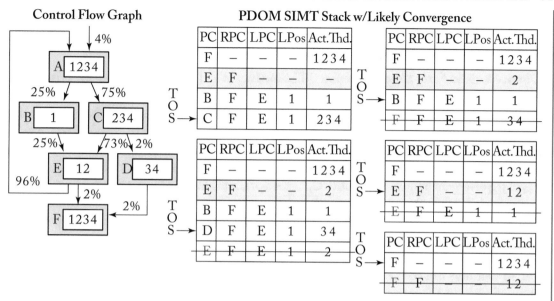

Figure 3.19: Earlier reconvergence points *before* the immediate post-dominator. Likely convergence point capture this earlier reconvergence at E.

reconvergence point, of the divergent branch, so that threads in this entry will reconverge to the definite reconvergence point.

As a warp executes with the top entry in the SIMT stack, it compares its PC against both the RPC field (just as it does with baseline SIMT stack), as well as the LPC field. If PC == LPC, the SIMT stack is popped, and threads in this popped entry is merged into the LPC entry. Otherwise, if PC == RPC, the SIMT stack is simply popped—the RPC entry already records these threads in its active mask. When the LPC entry reaches the top of the SIMT stack, it is executed just like any other SIMT stack entries, or popped directly if its active mask is empty.

Thread Frontiers.   Diamos et al. [2011] depart from the SIMT stack altogether and instead propose to reconverged threads after divergence via *thread frontiers*. A compiler supporting thread frontiers sorts the basic blocks in a kernel according to their topological order. In this way, threads executing at an instruction at a higher PC can never jump to an instruction at a lower PC. Loops are handled by placing the loop exit at the end of the loop body. With this sorted code layout, a diverged warp will eventually reconverge by prioritizing threads with lower PCs (allowing them to catch up).

Compared to SIMT stacks with immediate post-dominator reconvergence, reconvergence via thread frontiers yields higher SIMD efficiency for applications with unstructured control flow. The evaluation semantics of multi-expression conditional statements and the use of exceptions can both generate code with unstructured control flow. SIMT stacks extended with likely convergence points can yield similar SIMD efficiency improvement on applications with un-structured control flow; however, each entry in the SIMT stack may only have a finite number of likely convergence points, whereas the thread frontier approach has no such restriction.

Oppportunistic Early Reconvergence.   ElTantaway et al. [2014] proposes *opportunistic early reconvergence* (OREC), boosting the SIMD efficiency for GPU applications with unstructured control flow without any additional compiler analysis. OREC builds upon the Multi-Path (MP) SIMT Stack introduced in the same paper (see Section 3.4.2). The MP SIMT Stack uses a separate warp-split table holding the current set of warp-splits available for execution. At a divergent branch, new warp-splits are created with the branch target PCs and the reconvengence PC of the divergent branch. With OREC, instead of simply inserting these new warp-splits into the warp-split table, the hardware searches through the warp-split table for an existing warp-split with the same starting PC and RPC. If such a warp-split exists, the hardware create an early reconvergence point in the Reconvergence Table to converge the two warp-splits before the original RPC. The early reconvergence point synchronizes the two warp-split at a particular PC, so that they can be merged even when the existing warp-split have advanced through the diverged path. In ElTantaway et al. [2014] the early reconvergence point is the next PC of the existing warp-split.

3.4.3   ADDING MIMD CAPABILITY

The following proposals improve GPUs' compatibility with divergent control flow by incorpo-rating some limited amount of MIMD capability. All of these proposals offer two modes of operation:

- a SIMD mode, where the front-end issues one instruction to be executed across all threads in a warp; or

- a MIMD mode, where the front-end issues different instructions for each thread in a diverged warp.

When a warp is not diverged, it executes in SIMD mode to capture the control-flow locality exhibited by the threads in the warp, with energy efficiency comparable to traditional SIMD architectures. The warp switches to MIMD mode when it diverges. The warp runs less efficiently in this mode, but the performance penalty is lower than the penalty on a traditional SIMD architecture.

Vector-Thread Architecture.   Vector-thread (VT) architecture [Krashinsky et al., 2004] com-bines aspects of both SIMD and MIMD architectures, with the goal of capturing the best of both approaches. A VT architecture features a set of lanes that are connected to a common L1 instruction cache. In SIMD mode, all lanes receive instructions directly from the L1 instruction cache for lockstep execution, but each lane may switch to a MIMD model, running at its own pace with instructions from its L0 cache. A recent comparison with traditional SIMT architectures (e.g., GPUs) by Lee et al. [2011] shows that VT architectures have comparable efficiency with regular parallel applications, while performing much more efficiently with irregular parallel applications.

Temporal SIMT.   Temporal SIMT [Keckler et al., 2011, Krashinsky, 2011] permits each lane to execute in MIMD fashion, similar to VT architecture. However, instead of running a warp across all lanes in lockstep, it time-multiplexes the execution of a warp through a single lane, and each lane runs a separate set warp. Temporal SIMT achieves the efficiency of SIMD hardware by fetching each instruction only once for the whole warp. This amortizes the control flow overhead across time, while the traditional SIMD architecture amortizes the same overhead across multiple lanes in space.

Variable Warp Size Architecture.   Variable Warp Size (VWS) Architecture [Rogers et al., 2015] contains multiple (e.g., 8) slices, each of which contains a fetch and decode unit, so that each slice may execute different instructions simultaneously, similar to VT and Temporal SIMT. Instead of time-multiplexing large warps via narrow data path, each slice in VWS consists of narrow (4-wide) warps. These narrow warps are then grouped into larger execution entities called *gangs*. Each gang contains a warp from each slice.

In applications with no branch divergence, warps in a gang executes in lock-steps, obtaining instructions from a shared fetch unit and a shared L1 instruction cache. Upon encountering a branch divergence (or a memory divergence), the gang splits into multiple gangs. The new gangs may further splits until the point where every warp is in its own gang. At that point, these single-warp gangs are executed on their own slices individually via the slices' fetch unit and private L0 instruction cache. These split gangs are merged back together into the original gang opportunistically via hardware that compares the PC of the individual gangs. If they all match, the original gang is recreated. Rogers et al. [2015] also proposed inserting a gang-level synchronization barrier at the immediate post-dominator of the first divergent branch.

This book also evaluated performance impact of the capacity of the L0 instruction cache in each slice, in relation to the shared L1 instruction cache bandwidth. In non-ganged mode, the L0 caches in the slices may all requests instructions from the L1 cache simultaneously, creating a bandwidth bottleneck. Their evaluation shows that even for divergent applications, a 256-Byte L0 cache can filter out most of the requests to the shared L1 cache, As a result, the L1 cache can cover most of the bandwidth deficiency with just 2× bandwidth of the baseline SIMT architecture.

Simultaneous Branch Interweaving.   Brunie et al. [2012] propose simultaneous branch and warp interweaving (SBI and SWI) after the publication of thread block compaction. They extend the GPU SIMT front-end to support issuing two different instructions per cycle. SBI co-issues instructions from the same warp when it encounters a branch divergence. Executing both targets of a divergence branch at the same time eliminates its performance penalty significantly.

3.4.4   COMPLEXITY-EFFECTIVE DIVERGENCE MANAGEMENT

The area requirement of the baseline SIMT stack for each warp is only 32×64 bits (or as low as 6×64 bits with the optimization used in AMD GCN). While this area is small in comparison to the register file in a SIMT core, this area scales directly with the number of in-flight warps in the GPU, as well as the number of threads per warp. Moreover, in typical GPU applications where branch divergence is rare, the SIMT stack occupies area that can otherwise be used to boost the application throughput in other ways. A number of proposals replace the SIMT stack with alternative mechanisms, which share resources that can be used in other ways when a warp does not encounter any branch divergence.

SIMT Stack in Scalar Register File.   AMD GCN [AMD, 2012] features a scalar register file that is shared by all threads in a warp. Its registers can be used as predication registers to control the activity of each thread in a warp. The compiler uses this scalar register file to emulate a SIMT stack in software when it detects potentially divergent branch in the compute kernel. The GCN architecture features special instructions to accelerate the SIMT stack emulation.

One optimization that minimizes the number of scalar registers required to support the worst-case divergence is to prioritize the execution of the target with the fewer number of active threads. This allows the worst-case divergence to be supported with $\log_2(\#threads\ per\ warp)$ scalar registers, which is far fewer than the entries required by the baseline SIMT stack. Furthermore, when the compute kernel has no potentially divergent branch, the compiler can use the scalar registers reserved for the SIMT stack for other scalar computation.

Thread Frontiers.   As mentioned in Section 3.4.2, Diamos et al. [2011] replaces the SIMT stack with *thread frontiers*. With thread frontiers, each thread maintains its own PC in the register file, and the code is topologically sorted so that reconvergence point of a branch always have a higher PC. When a warp diverges, it always prioritize the threads with the lowest PC among all of its threads. This set of threads is known as the thread frontier of the warp. Prioritizing the execution of the threads in the frontier implicitly forces all threads further ahead in the program to wait at the reconvergence point of the branch to be merged.

Since the per-thread PCs are only needed when the compute kernel contains potentially divergent branches, the compiler only needs to allocate a PC register in these compute kernels. In other compute kernels, the extra register storage works to improve the warp occupancy, increasing the number of warps that each SIMT core can sustain to better tolerate memory latency.

Stackless SIMT.   Asanovic et al. [2013] proposes extending the temporal SIMT architecture with a `syncwarp` instruction. In this proposal, threads in a warp run in lock-step when they execute in the *convergent regions* of the compute kernel, where the compiler guarantees that the warp can never diverge. At a divergent branch, each thread in a warp follows its own control-flow path with its private PC, leveraging the MIMD capability in the temporal SIMT architecture.

The compiler places a `syncwarp` instruction at the reconvergence point of the divergent branch. This forces all threads in the diverged warp to sync up at the reconvergence point before entering another convergent region of the compute kernel.

While this mechanism does not capture reconvergence possible with nested divergent branches, it is nevertheless a cheaper alternative that can provide comparable performance to the baseline SIMT stack for GPU applications that rarely exhibit branch divergence. The paper introduces a combined convergent and variant analysis that allows the compiler to determine operations in an arbitrary compute kernel that is eligible for *scalarization* and/or *affine transformation*. In the context of stackless SIMT, the same analysis allows the compiler to determine the convergent and divergent regions in an arbitrary compute kernel.

1. The compiler first assumes all basic-blocks to be *thread-invariant*.

2. It marks all instructions dependent on the thread ID, atomic instructions, and memory instructions on volatile memory as *thread-variant*.

3. It then iteratively marks all instruction dependent on thread-variant instructions as thread-variant as well.

4. All instructions in basic-blocks that are *control dependent* on a thread-variant branch instruction are also thread-variant. Essentially, instructions beyond the immediate post-dominator of a thread-variant branch may remain thread-invariant, as long as they are not marked as thread-variant for other conditions.

This analysis allows the compiler to detect branches that are taken uniformly by all threads in each warp. Since these branches do not cause warp to diverge, the compiler does not need to insert code to detect dynamic divergence for these branches, nor does it need to insert `syncwarp` instructions at their immediate post-dominators to enforce reconvergence.

Predication.   Before incorporating a full SIMT stack into the architecture, GPUs with programmable shaders has been supporting limited control-flow constructs in the shader programs via *predications*, just like traditional vector processors. Prediction remains in modern GPUs as a low-overhead way to handle simple if-branches, avoiding the overhead of pushing and popping the SIMT stack. In NVIDIA's implementation, each instruction is extended with an extra operand field to specify its predication registers. Predication registers are essentially scalar registers dedicated to control flow.

Lee et al. [2014b] proposes a *thread-aware prediction algorithm* that extends the application of predication to arbitrary control-flow, with performance comparable to the NVIDIA's SIMT stack. The thread-aware predication algorithm extends the standard Control-Flow Dependency Graph (CDG) with predication nodes at each branch. The prediction required by each basic block can then be computed based on their control-flow dependency, and further optimized rigorously without breaking the functional behavior. The paper then describes two optimizations based on this thread-aware CDG, and the convergence and variance analysis in their prior work [Asanovic et al., 2013].

- *Static branch-uniformity optimization* is applied when the compiler can guarantee that the branch can be taken uniformly across the entire warp, as deduced by the convergence analysis. In this case, the compiler can replace the predication generation with a uniform branch instruction.

- *Runtime branch-uniformity optimization* is applied in other cases. The compiler issue consensual branches (`cbranch.ifnone`) that are only taken when given a null predicate (i.e., all threads disabled). This allows the warp to skip through code with null predicates—a key advantage offered by SIMT stacks. This approach differs from prior efforts for vector processors, such as BOSCC, in that it relies on structure analysis to determine the candidates for this optimization.

While both predication and SIMT stacks are fundamentally providing the same functionality at a similar energy and area cost, Lee et al. [2014b] highlighted the following trade-offs between the two approaches.

- Since different branch targets are guarded by different predication registers, the compiler can schedule instructions from different branch targets, interleaving the execution of different branch targets to exploit thread-level parallelism (TLP) that would otherwise require more advanced hardware branch divergence management.

- Predication tends to increase register pressure, which in turn reduces warp occupancy and imposes an overall performance penalty. This happens because a conservative register allocation cannot reuse registers for both sides of a branch. It cannot robustly prove that instructions from different branch targets are operating no exclusive sets of lanes in the registers. The uniform and consensual branch instructions inserted by the two proposed optimizations alleviate this problem.

- Predication may impact the dynamic instruction count in multiple ways. In some cases, the overhead of checking uniform branches increases the dynamic instruction count significantly. Alternatively, not performing the check means that some paths are executed with a null predication mask. In other cases, it removes push/pop instructions required to maintain the SIMT stack.

In the end, the paper proposes new instructions to reduce the overhead of predication.

- For function calls and indirect branches, they propose a new `find_unique` instruction to serially execute each branch target/function via a loop.

- `cbranch.ifany` (in addition to existing consensual branch instructions `cbranch.ifnone` and `cbranch.ifall`) will help reduce the instruction count overhead introduced by dynamic uniform branch detection.

3.5   RESEARCH DIRECTIONS ON SCALARIZATION AND AFFINE EXECUTION

As described in Chapter 2, GPU computing APIs, such as CUDA and OpenCL, feature a MIMD-like programming model that allows the programmer to launch a large array of scalar threads onto the GPU. While each of these scalar threads can follow its unique execution path and may access arbitrary memory locations, in the common case, they all follow a small set of execution paths and perform similar operations. The convergent control-flow among GPU threads is exploited on most, if not all, modern GPUs via the SIMT execution model, where scalar threads are grouped into warps that runs on SIMD hardware (see Section 3.1.1).

This section summarizes a series of research that further exploit the similarity of these scalar threads via *scalarization* and *affine execution*. The key insight of these research lies in the observation of *value structure* [Kim et al., 2013] across threads executing the same compute kernel. The two types of value structure, *uniform* and *affine*, are illustrated the compute kernel in Example 3.3.

Uniform Variable A variable that has the same constant value for every thread in the compute kernel. In Algorithm 3.3, the variable a, as well as the literals THRESHOLD and Y_MAX_VALUE, all have uniform value across all threads in the compute kernel. A uniform variable can be stored in a single scalar register, and reused by all threads in the compute kernel.

Affine Variable A variable with values that is a linear function of thread ID for every thread in the compute kernel. In Algorithm 3.3, the memory address of the variable y[idx] can be represented as an *affine* transform of the thread ID threadIdx.x:

```
&(y[idx]) = &(y[0]) + size(int) * threadIdx.x;
```

This affine representation can be stored as a pair of scalar values, a *base* and a *stride*, which is far more compact than the fully expanded vector.

There are multiple research proposal on how to *detect* and *exploit* uniform or affine variables in GPUs. The rest of this section summarizes these proposals in these two aspects.

3.5.1   DETECTION OF UNIFORM OR AFFINE VARIABLES

There are two main approaches to detect the existence of uniform or affine variables in a GPU compute kernel: Compiler-Driven Detection and Detection via Hardware.

Compiler-Driven Detection

One way to detect the existence of uniform or affine variables in a GPU compute kernel is to do so via a special compiler analysis. This is possible because the existing GPU programming models, CUDA and OpenCL, already provides means for the programmer to declare a variable as

Algorithm 3.3 Example of scalar and affine operations in a compute kernel (from [Kim et al., 2013]).

```
__global__ void vsadd( int y[], int a )
{
  int idx = threadIdx.x;
  y[idx] = y[idx] + a;
  if ( y[idx] > THRESHOLD )
    y[idx] = Y_MAX_VALUE;
}
```

constant through out the compute kernel, as well as providing special variable for the thread ID. The compiler can perform a control-dependency analysis to detect variables that are dependent purely on constants and thread IDs, and mark them as uniform/affine. Operations that work solely on uniform/affine variables are then candidates for *scalarization*.

AMD GCN [AMD, 2012] relies on the compiler to detect uniform variables and scalar operations that can be stored and processed by a dedicated scalar processor.

Asanovic et al. [2013] introduce a combined convergent and variant analysis that allows the compiler to determine operations in an arbitrary compute kernel that is eligible for *scalarization* and/or *affine transformation*. Instructions within the convergent regions of a compute kernel can be converted into scalar/affine instructions. At any transition from divergent to convergent regions of a compute kernel, the compiler inserts a syncwarp instruction to handle control-flow induced register dependencies between the two regions. Asanovic et al. [2013] adopted this analysis to generate scalar operations for the Temporal-SIMT architecture [Keckler et al., 2011, Krashinsky, 2011].

Decoupled Affine Computation (DAC) [Wang and Lin, 2017] relies on a similar compiler analysis to extract scalar and affine candidates to be decoupled into a separate warp. Wang and Lin [2017] augments the process with a divergent affine analysis, with the goal to extract strands of instructions that has been affine from the start of the compute kernel. These strands of affine instructions are decoupled from the main kernel into an affine kernel that feeds data into the main kernel via a hardware queue.

Hardware Detection

Detecting uniform/affine variables in hardware offers two potential advantage over compiler-driven detection.

1. This allows scalarization and affine execution to be applied with the original GPU instruction set architecture. It saves the effort to co-develop a special scalarization compiler along with the hardware.

2. The hardware detection happens during the compute kernel execution. As a result, it is capable of detecting uniform/affine variables that occurs dynamically, but are missed by the static analysis.

Tag-Based Detection.   Collange et al. [2010] introduce a tag-based detection system. In this system, each GPU register is extended with a tag, indicating if the register contains uniform, affine, or generic vector values. At the launch of a compute kernel, the tag of the register that contains the thread ID is set to affine state. Instructions that broadcast values from a single location in constant or shared memory set the tag of the destination register to the uniform state. During kernel execution, states of the registers are propagated across arithmetic instruction from source to destination operands according to simple rules in Table 3.1. While this tag-based detection has little hardware overhead, it tends to be conservative—for example, it conservatively rules the output of multiplication between uniform and affine variables as vector variable.

Table 3.1: Examples of rules of uniform and affine state propagation across instructions from Collange et al. [2010]. For each operation, the first row and column shows states of the input operands, and the remaining entries show the state of the output operand for every permutation of input operand states (U = uniform, A = affine, V = vector).

+	U	A	V	×	U	A	V	<<	U	A	V
U	U	A	V	U	U	V	V	U	U	A	V
A	A	V	V	A	V	V	V	A	V	V	V
V	V	V	V	V	V	V	V	V	V	V	V

FG-SIMT architecture [Kim et al., 2013] extends the tag-based detection mechanism from Collange et al. [2010] with better support for branches. Affine branches, or branches that compares between affine operands, is resolved via the scalar datapath if one of the operand is a uniform. Kim et al. [2013] also introduce a *lazy expansion* scheme, where affine registers are lazily expanded into full vector registers after a divergent branch or a predicated instruction. This expansion is required to allow a subset of threads in a divergent warp to update their slots in the destination register, while leaving other slots unchanged—this maintains the SIMT execution semantics. In contrast to a more naive, eager, expansion scheme that expands every affine register after the first divergent branch, the lazy expansion scheme eliminates many unnecessary expansion.

Comparison at Write-Back.   Gilani et al. [2013] introduce a more aggressive mechanism to detect uniform variables by comparing the register values from all threads in a warp at each write-back of a vector instruction. At the detection of an uniform variable, the detection logic reroutes the write-back to a scalar register file, and updates an internal table to remember the state of the register. Subsequent use of the register is then redirected to the scalar register file. Instructions with all operands from the scalar register file are executed on a separate scalar pipeline.

Lee et al. [2015] use a similar detection scheme. Instead of a simple uniform detector, they augment the register write-back stage with a register value compressor that transforms an incoming vector of values into a tuple of $< base, delta, immediate >$ (BDI) using the algorithm introduced by Pekhimenko et al. [2012].

Wong et al. [2016] introduce *Warp Approximation*, a framework to exploit approximate computing within a warp which also features a detection at register write-back. The detector computes the smallest *d-similarity*, two given values sharing *d*-MSBs, among all values in the vector written back to the register file. The register with a higher than threshold d-similarity is marked as *similar*, which is then used to determine the eligibility of approximation execution in subsequent dependent instructions.

Like the proposal from Lee et al. [2015], G-Scalar [Liu et al., 2017] features a register value compressor at the register write-back stage, but the compressor employs a simpler algorithm that only extracts common bytes used by all values across all lanes in the warp. If all bytes are common, the register contains a uniform variable. Any instruction that operates only on uniform variables can be scalarized.

G-Scalar also extended the register value compressor to detect operation that is eligible for scalar execution under branch divergence. All prior proposals revert back to vector execution as soon as the warp diverge. Liu et al. [2017] observe that in many instruction under branch divergence, the operand values of the active lanes are uniform. Instructions using these partially uniform registers are actually eligible for scalar execution. They then extend the register value compressor to check only values from the active lanes using a special logic. This greatly increases the number of scalar instructions across various GPU compute workloads. Note that under divergence, the written register is not compressed.

3.5.2   EXPLOITING UNIFORM OR AFFINE VARIABLES IN GPU

The design of a GPU may exploit the existence of value structure in compute kernels in multiple ways.

Compressed Register Storage

The compact representation of uniform and affine variables allows them to be stored in the Register File with far fewer bits. The salvaged storage can be used to sustain more inflight warps, increasing a GPU's tolerance to memory latency with the same register file resource.

Scalar Register File.   Many proposals/designs exploit uniform or affine variables in GPU features a dedicated register file for scalar/affine values.

- AMD GCN architecture features a scalar register file that is accessible by both scalar and vector pipelines.

- FG-SIMT architecture [Kim et al., 2013] stores uniform/affine values in a separate Affine SIMT Register File (ASRF). The ASRF records the state (affine/uniform/vector) of each register, allowing the control logic to detect operations eligible for direction execution on the Control Processor.

- The dynamic uniform detection proposal from Gilani et al. [2013] stores the dynamically detected uniform values into a dedicated scalar register file.

Partial Register File Access.   Lee et al. [2015] apply base, delta, immediate (BDI) compression to registers written back to the register file. The compressed registers are decompressed back to normal vectors as it is read back as a source operand. In this scheme, each compressed register still occupies the same storage slot as an uncompressed register, but only a subset of the register banks, thus it takes less energy to read out the compressed representation of the register.

Warp Approximate architecture [Wong et al., 2016] reduces the register read/write energy use by only accessing the lane corresponding to the representative thread selected via similarity detection.

Similarly, G-Scalar [Liu et al., 2017] features compressed registers that occupy only a subset of banks allocated or the uncompressed register to reduce the energy for register read.

Dedicated Affine Warp.   Decoupled Affine Computation (DAC) [Wang and Lin, 2017] buffers all compiler-extracted affine variables in the registers of a dedicated affine warp. This affine warp shares the same vector register file storage just as the rest of the non-affine warps, but the affine warp uses individual lanes of each individual register entry to store the base, and the deltas for different non-affine warps.

Scalarize Operations

Aside from efficient storage, the operations with uniform or affine variables can be *scalarized*. Instead of repeating the same operation across all threads in a warp via a SIMD datapath, a scalar operation can be done once in single scalar datapath, consuming far less energy in the process. In general, an arithmetic operation can be scalarized if its input operands consist only of uniform or affine variables.

Dedicated Scalar Pipeline.   AMD's GCN architecture features a dedicated scalar pipeline that executes scalar instructions generated by the compiler. FG-SIMT architecture [Kim et al., 2013] features a control processor that is capable of executing dynamically detected affine operations directly without evoking the SIMD datapath.

In both implementations, the scalar pipeline also handles the control flow and predication of the SIMD pipeline. The decoupling means that many system-related features (for example, communication with host processor) can also be offloaded to the scalar pipeline, freeing the SIMD datapath from the burden of implementing the full instruction set.

Clock-Gated SIMD Datapath.   Warp Approximate architecture [Wong et al., 2016] and G-Scalar [Liu et al., 2017] both executes dynamically detected scalar instructions on one of the lanes in the SIMD datapath. When this happens, other lanes are clock-gated to reduce dynamic power consumption.

This approach eliminates the duplicating effort of supporting the full instruction set on the dedicated scalar datapath, or having to triage the subset to be implemented on the scalar datapath. For example, G-Scalar [Liu et al., 2017] can scalarize instructions supported by the special function units with relatively low overhead.

Aggregate to Affine Warp.   Decoupled Affine Computation (DAC) [Wang and Lin, 2017] aggregates the affine operations from multiple warps into a single affine warp per SIMT core. This affine warp executes on the SIMD datapath just as other warps, but each instruction executed operates simultaneously on the affine representation of multiple warps.

Memory Access Acceleration

When a uniform or affine variable is used to represent the address of a memory operation (load-/store), the memory locations touched by the memory operation is highly predictable—each successive location is separate by a known stride. This allows for various optimizations. For example, the memory coalescing of memory locations with a known stride is far simpler than coalescing of arbitrary random locations. The affine variable can also be used to represent a bulk transfer with a single instruction rather than via loops of load/store instructions.

FG-SIMT architecture [Kim et al., 2013] features a special address generation unit in the control process to expand memory accesses with affine addresses into the actual addresses. Since affine addresses have a fixed stride between threads, coalescing these affine memory accesses into cache lines can be done with simpler hardware.

Decoupled Affine Computation (DAC) [Wang and Lin, 2017] also features similar optimizations to exploit fixed strides in affine memory accesses. In addition, it uses an affine warp to execute ahead of the rest of the non-affine warps, prefetching data for these warps. The prefetched data is stored in the L1 cache, retrieved later by the corresponding non-affine warp via a special dequeue instruction.

3.6   RESEARCH DIRECTIONS ON REGISTER FILE ARCHITECTURE

Modern GPUs employ a large number of hardware threads (warps), multiplexing their execution on a far fewer (still large) number of ALUs, to tolerate both pipeline and memory access latencies. To allow fast and efficient switching between warps, GPUs use hardware warp schedulers and store the registers of all hardware threads in on-chip register files. On many GPU architectures, the capacity of these register files is substantial, and sometimes exceeding the capacity of the last-level cache, due to the wide-SIMD datapaths used in GPU, as well as the shear number of warps required to tolerate hundreds of cycles of memory access latency. For example,

NVIDIA's Fermi GPU can sustain over 20,000 in-flight threads, and has an aggregate register capacity of 2 MB.

To minimize the area consumed by register file storage, register files on GPUs are generally implemented via low-port count SRAM banks. The SRAM banks are accessed in parallel to supply the operand bandwidth required to sustain instructions running on the wide-SIMD pipeline at peak throughput. As described earlier in this chapter, some GPUs use an operand collector to orchestrate operand accesses from multiple instructions to minimize bank-conflict penalties.

Accessing these large register files consumes a high amount of dynamic energy at each access, and their large size also leads to high static power consumption. On a NVIDIA GTX280 GPU, nearly 10% of the total GPU power is consumed by the register file. This provides clear incentives to innovate on GPU register file architectures to reduce their energy consumption. As a result, there have been a large number of research papers on this topic in recent years. The rest of this section summarizes several research proposals aimed to achieve this goal.

3.6.1   HIERARCHICAL REGISTER FILE

Gebhart et al. [2011b] observe that among a set of real-world graphics and compute workloads, up to 70% of the values produced by an instruction are only read once, and only 10% are only read more than twice. To capture this short lifetime among a majority of the register values, they propose extending the main register file on GPU with a *register file cache* (RFC). This forms a hierarchy of the register file, and dramatically reduces the access frequency to the main register file.

In this work, the RFC allocates a new entry, via a FIFO replacement policy, for the destination operand of every instruction. Source operands that miss the RFC are not loaded onto the RFC to reduce pollution of the already small RFC. By default, every value evicted from RFC is written back to the main register file. However, since many of these values are never read again, Gebhart et al. [2011b] extend the hardware-only RFC with compile time-generated static liveness information. An extra bit is added to the instruction encoding to indicate the last instruction consuming a register value. A register that has been read for the last time is marked dead in the RFC. At eviction, it will not be written back to the main register file.

To further reduce the size of the RFC, Gebhart et al. [2011b] combine it with a two-level warp scheduler. This two-level warp scheduler restricts execution to a pool of *active* warps, which only consists of a small subset of the warps in each SIMT core. This work considers an active warp pool of 4–8 warps, out of 32 warps in total for each SIMT core. The RFC only holds values from the active warps, and is therefore smaller. A warp is removed from the active pool at long-latency operations, such as global memory loads or texture fetches. When this occurs, the RFC entries of the warp are flushed, freeing up space for a different warp made active by the second-level scheduler.

Compile-Time Managed Register File Hierarchy.   Gebhart et al. [2011a] further extend this register file hierarchy with a Last Result File (LRF), which just buffers the register value produced by the last instruction of each active warp. This work also replaces a hardware-managed RFC with a compile-time managed operand register file (ORF). Movements of values in and out of the ORF is managed explicitly by the compiler. This removes the tag-lookup required by the RFC. The compiler also has a more holistic view of the register usage pattern in most GPU workloads, allowing it to make more optimal decisions. This work also extends the two-level warp scheduler so that the compiler indicates when a warp can be switched out of the active pool. This is required to coordinate the content of the ORF with the activeness of the warp, moving all live data from the ORF back to the main register file before the warp is switched out.

3.6.2   DROWSY STATE REGISTER FILE

Abdel-Majeed and Annavaram [2013] proposes a tri-modal register file design that reduces the leakage power of the large GPU register file. Each entry in the tri-modal register file can switch between ON, OFF, and Drowsy mode. ON mode is the normal mode of operation; OFF mode does not retain the value of the register; Drowsy mode retains the value of the register, but the entry needs to be awaken to ON mode before access. In this work, all non-allocated registers are in OFF mode, and all allocated registers are placed into drowsy state immediately after each access. This policy takes advantage of the long delays between consecutive access to the same register on GPU, due to the fine-grained multithreading on GPU, to allow registers in the register file to spend most time in drowsy mode. The long pipeline in GPU also means that additional latency of waking up a register from drowsy state does not introduce significant performance penalty.

3.6.3   REGISTER FILE VIRTUALIZATION

Tarjan and Skadron [2011] observe that while waiting for memory operations the number of live registers in a GPU thread tends to be small. For some GPU applications they claim up to 60% of registers go unused. They propose reducing the size of the physical register file by up to 50% or double the number of concurrently executing threads by using register renaming to virtualize physical registers. In the proposed mechanism, a thread starts executing with no registers allocated and physical registers are allocated to destination registers as instructions are decoded. Tarjan and Skadron further suggest that deallocation of physical registers can be enhanced by employing compiler analysis to determine the last read of a register. They propose "final read annotations" and suggest adding "a bit for each operand to indicate whether it is a last read" and point out this may require additional bits in the instruction encoding.

Jeon et al. [2015] quantify the impact of reducing the GPU register file size by spilling registers to memory. They find that reducing the size of the register file by 50% by employing spilling increased execution time by an average of 73%. They review older proposals for reclaiming physical registers early when employing register renaming on CPUs employing out-of-order execution. The propose addressing the issue of the additional bits required to add "final read annotations" by adding "metadata instructions" that efficiently encode when physical registers can be reclaimed and generate these using register lifetime liveness analysis. An important observation they make is that branch divergence must be taken into account when determining where it is safe to reclaim physical registers (further elaborated upon by Kloosterman et al. [2017]). For a 128 KB register file a straightforward implementation of Jeon et al.'s renaming technique requires 3.8 KB of renaming hardware. They show this overhead can be reduced to 1 KB by not renaming registers with long lifetimes. To exploit this opportunity they propose using renaming only for registers with logical register numbers larger than a compiler determined threshold. Jeon et al. further propose using renaming to enable power gating of register file subarrays. They evaluate the effectiveness of their detailed proposal for supporting register file virtualization via register renaming showing that a reduction in the size of the register file by 50% with no performance loss is indeed obtainable.

3.6.4   PARTITIONED REGISTER FILE

Abdel-Majeed et al. [2017] introduces the *Pilot Register File*, which partitions the GPU register file into a fast and a slow register file (FRF and SRF). The FRF is implemented using regular SRAMs, whereas the SRF is implemented using near-threshold voltage (NTV) SRAMs. Compared to regular SRAMs, NTV SRAMs feature far lower access energy as well as a much lower leakage power. In exchange, access latency to the NTV SRAMs are far slower, often consists of several cycles (instead of one cycle in regular SRAMs). In this work, the SRF is significantly larger than the FRF. Every warp has 4 entries in the FRF. The key is to use FRF to service most of the access to compensate for the slowness of the SRF. The additional latency for accessing the SRF is handled by the operand collector. The FRF is further enhanced with a low power mode using FinFET's back gate control. This allow the FRF of an inactive warp to switch to low power mode. This allows FRF to reap the benefit of two-level scheduler without explicit scheduling warps in and out of an active pool.

This work is distinct from the hierarchical register file in that the different partitions hold an exclusive set of registers, and the partition remains constant throughout the lifetime of the warp. Instead of using a compiler to determine the set of registers to be placed in the FRF, Abdel-Majeed et al. [2017] employs a pilot CTA at each kernel launch to profile the most-frequently used registers. This set of high-use registers are recorded in a lookup table that is accessible to every subsequent warp from the kernel launch.

3.6.5   REGLESS

Kloosterman et al. [2017] introduces *RegLess*, which aims at eliminating the register file and replace it with an operand staging buffer. The paper observes that over relatively short spans of time the number of registers accessed is a small fraction of the total register file capacity. For example, over a period of 100 cycles many of the applications they evaluated accessed less than

10% of a 2048 KB register file when using a GTO or two-level warp scheduler. To take advantage of this observation RegLess uses a compiler algorithm to divide up kernel execution into regions. Regions are contiguous instructions within a single basic block. The boundary between regions is selected so as to limit the number of live registers. Using the region annotations, a Capacity Manager (CM) determines which warps are eligible for scheduling. When a warp begins executing instructions from a new region the registers used in that region are brought into an Operand Staging Unit (OSU) from a backing storage area allocated in global memory and potentially cached in the L1 data cache. The OSU is essentially a cache consisting of eight banks which provides enough bandwidth to service two instructions per cycle. To avoid stalling while accessing data in the OSU, the CM preloads registers before issuing the first instruction in a region. To manage the preloading process the CM maintains a state machine for each warp indicating whether the registers needed for the next region are present in the OSU. To reduce the amount of memory traffic generated between the OSU and memory hierarchy, RegLess employs register compression techniques that exploit affine values (see Section 3.5) are employed.

Kloosterman et al. performed a detailed evaluation of their proposal including Verilog synthesis and extraction of parasitic capacitance and resistance values of the hardware units introduced by RegLess. Their evaluation shows a 512 entry OSU can achieves slightly better performance versus a 2048 KB register file while occupying only 25% of the space and reducing overall GPU energy consumption by 11%.

CHAPTER 4

Memory System

This chapter explores the memory system of GPUs. GPU computing kernels interact with the memory system through load and store instructions. Traditional graphics applications interact with several memory spaces such as texture, constant, and render surfaces. While access to these memory spaces is available in GPGPU programming APIs like CUDA, we will focus on memory spaces employed in GPGPU programming in this chapter and in particular the microarchitecture support required to implement them.

CPUs typically include two separate memory spaces: The register file and memory. Modern GPUs logically subdivide memory further into local and global memory spaces. The local memory space is private per thread and typically used for register spilling while global memory is used for data structures that are shared among multiple threads. In addition, modern GPUs typically implement a programmer managed scratchpad memory with shared access among threads that execute together in a cooperative thread array. One motivation for including a shared address space is that in many applications a programmer knows which data needs to be accessed at a given step in a computation. By loading all of this data into shared memory at once they can overlap long latency off-chip memory accesses and avoid long latency accesses to memory while performing computation on this data. More importantly, the number of bytes that can be transferred between the GPU and off-chip memory in a given amount of time (DRAM bandwidth) is small relative to the number of instructions that can be executed in that same amount of time. Moreover, the energy consumed to transfer data between off-chip memory and the GPU is orders of magnitude higher than the energy consumed accessing data from on-chip memory. Thus, accessing data from on-chip memory yields higher performance and saves energy.

We divide our discussion of the memory system into two parts reflecting the division of memory into portions that reside within the GPU cores and within memory partitions that connect to off-chip DRAM chips.

4.1 FIRST-LEVEL MEMORY STRUCTURES

This section describes the first-level cache structures found on GPUs with a focus on the unified L1 data cache and scratch pad "shared memory" and how these interact with the core pipeline. We also include a brief discussion of a typical microarchitecture for an L1 texture cache. We include discussion of the texture cache, which has found limited use in GPU computing applications, as it provides some insights and intuition as to how GPUs differ from CPUs. A recent patent describes how one might unify the texture cache and L1 data (e.g., as found in NVIDIA's

Maxwell and Pascal GPUs) [Heinrich et al., 2017]. We defer discussion of this design until after first considering how texture caches are organized. An interesting aspect of the first-level memory structures in GPUs is how they interact with the core pipeline when hazards are encountered. As noted in Chapter 3, pipeline hazards can be handled by replaying instructions. We expand on our earlier discussion of replay in this chapter with a focus on hazards in the memory system.

4.1.1   SCRATCHPAD MEMORY AND L1 DATA CACHE

In the CUDA programming model, "shared memory" refers to a relatively small memory space that is expected to have low latency but which is accessible to all threads within a given CTA. In other architectures, such a memory space is sometimes referred to as a scratchpad memory [Hofstee, 2005]. The latency to access this memory space is typically comparable to register file access latency. Indeed, early NVIDIA patents refer to CUDA "shared memory" as a Global Register File [Acocella and Goudy, 2010]. In OpenCL this memory space is referred to as "local memory." From a programmer perspective a key aspect to consider when using shared memory, beyond its limited capacity, is the potential for *bank conflicts*. The shared memory is implemented as a static random access memory (SRAM) and is described in some patents [Minkin et al., 2012] as being implemented with one bank per lane with each bank having one read port and one write port. Each thread has access to all of the banks. A *bank conflict* arises when more than one thread accesses the same bank on a given cycle and the threads wish to access distinct locations in that bank. Before considering in detail how the shared memory is implemented we first look at the L1 data cache.

The L1 data cache maintains a subset of the global memory address space in the cache. In some architectures the L1 cache contains only locations that are not modified by kernels, which helps avoid complications due to the lack of cache coherence on GPUs. From a programmer perspective a key consideration when accessing global memory is the relationship, with respect to each other, of memory locations accessed by different threads within a given warp. If all threads in a warp access locations that fall within a single L1 data cache block and that block is not present in the cache, then only a single request needs to be sent to lower level caches. Such accesses are said to be "coalesced." If the threads within a warp access different cache blocks then multiple memory accesses need to be generated. Such accesses are said to be uncoalesced. Programmers try to avoid both bank conflicts and uncoalesced accesses, but to ease programming the hardware allows both.

Figure 4.1 illustrates a GPU cache organization like that described by Minkin et al. [2012]. The design pictured implements a unified shared memory and L1 data cache, which is a feature introduced in NVIDIA's Fermi architecture that is also present in the Kepler architecture. At the center of the diagram is an SRAM data array ❺ which can be configured [Minkin et al., 2013] partly for direct mapped access for shared memory and partly as a set associative cache. The design supports a non-stalling interface with the instruction pipeline by using a replay mechanism when handling bank conflicts and L1 data cache misses. To help explain the operation of this cache architecture we first consider how shared memory accesses are processed, then consider coalesced cache hits, and finally consider cache misses and uncoalesced accesses. For all cases, a memory access request is first sent from the load/store unit inside the instruction pipeline to the L1 cache ❶. A memory access request consists of a set of memory addresses, one for each thread in a warp along with the operation type.

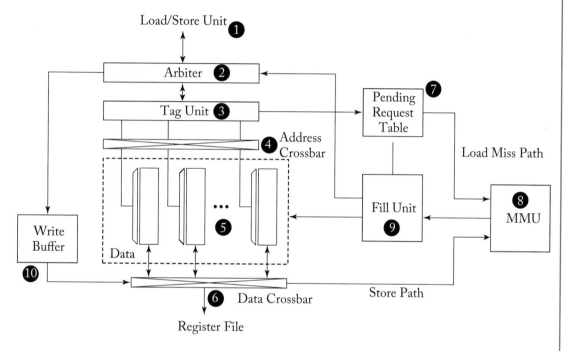

Figure 4.1: Unified L1 data cache and shared memory [Minkin et al., 2012].

Shared Memory Access Operations

For a shared memory accesses the arbiter determines whether the requested addresses within the warp will cause bank conflicts. If the requested addresses would cause one or more bank conflicts, the arbiter splits the request into two parts. The first part includes addresses for a subset of threads in the warp which do not have bank conflicts. This part of the original request is accepted by the arbiter for further processing by the cache. The second part contains those addresses that cause bank conflicts with addresses in the first part. This part of the original request is returned to the instruction pipeline and must be executed again. This subsequent execution is known as a "replay." There is a tradeoff in where the replay part of the original shared memory request is stored. While area can be saved by replaying the memory access instruction from the instruction buffer this consumes energy in accessing the large register file. A better alternative for energy efficiency may be to provide limited buffering for replaying memory access instructions in the load/store unit and avoiding scheduling memory access operations from the instruction buffer when free space in this buffer beings to run out. Before considering what happens to the replay request, let us consider how the accepted portion of the memory request is processed.

The accepted portion of a shared memory request bypasses tag lookup inside the tag unit ❸ as shared memory is direct mapped. When accepting a shared memory load request the arbiter schedules a writeback event to the register file inside the instruction pipeline as the latency of the direct mapped memory lookup is constant in the absence of bank conflicts. The tag unit determines which bank each thread's request maps to so as to control the address crossbar ❹ which distributes addresses to the individual banks within the data array. Each bank inside the data array ❺ is 32-bits wide and has its own decoder allowing for independent access to different rows in each bank. The data is returned to the appropriate thread's lane for storage in the register file via the data crossbar ❻. Only lanes corresponding to active threads in the warp write a value to the register file.

Assuming a single-cycle latency for shared memory lookup, the replayed portion of a shared memory request can access the L1 cache arbiter the cycle after the previous accepted portion. If this replayed portion encounters bank conflicts it is further subdivided into an accepted and replayed portion.

Cache Read Operations

Next, let us consider how a load to the global memory space is processed. As only a subset of the global memory space is cached in the L1 the tag unit will need to check whether the data is present in the cache or not. While the data array is highly banked to enable flexible access to shared memory by individual warps, access to global memory is restricted to a single cache block per cycle. This restriction helps to reduce tag storage overhead relative to the amount of cached data and is also a consequence of the standard interface to standard DRAM chips. The L1 cache block size is 128 bytes in Fermi and Kepler and is further divided into four 32-byte sectors [Liptay, 1968] in Maxwell and Pascal [NVIDIA Corp.]. The 32-byte sector size corresponds to the minimum size of data that can be read from a recent graphics DRAM chip in a single access (e.g., GDDR5). Each 128-byte cache block is composed of 32-bit entries at the same row in each of the 32 banks.

The load/store unit ❶ computes memory addresses and applies the coalescing rules to break a warp's memory access into individual coalesced accesses which are then fed into the arbiter ❷. The arbiter may reject a request if enough resources are not available. For example, if all ways in the cache set that the access maps to are busy or there are no free entries in the pending request table ❼, which is described below. Assuming enough resources are available to handle a miss, the arbiter requests the instruction pipeline schedule a writeback to the register file a fixed number of cycles in the future corresponding to a cache hit. In parallel the arbiter also requests the Tag Unit ❸ check whether the access in fact leads to a cache hit or a miss. In the event of a cache hit, the appropriate row of the data array ❺ is accessed in all banks and the data is returned ❻ to the register file in the instruction pipeline. As in the case of shared memory accesses, only register lanes corresponding to active threads are updated.

When accessing the Tag Unit, if it is determined that a request triggers a cache miss, the arbiter informs the load/store unit it must replay the request and in parallel it sends the request information to the pending request table (PRT) ❼. The pending request table provides functionality not unlike that supported by traditional miss-status holding registers [Kroft, 1981] in CPU cache memory systems. There are at least two versions of the pending request table described in NVIDIA patents [Minkin et al., 2012, Nyland et al., 2011]. The version associated with the L1 cache architecture shown in Figure 4.1 appears somewhat similar to a traditional MSHR. Traditional MSHRs for data caches contain the block address of a cache miss along with information on the block offset and associated register that needs to be written when the block is filled into the cache. Multiple misses to the same block are supported by recording multiple block offsets and registers. The PRT in Figure 4.1 supports merging two requests to the same block and records information needed to inform the instruction pipeline which deferred memory access to replay.

The L1 data cache shown in Figure 4.1 is virtually indexed and virtually tagged. This may be surprising when contrasted with modern CPU microarchitectures which mostly employ virtual indexed/physically tagged L1 data caches. CPUs use this organization to avoid the overheads of flushing the L1 data cache on context switches [Hennessy and Patterson, 2011]. While GPUs effectively perform a context switch every cycle that a warp issues, the warps are part of the same application. Page-based virtual memory is still advantageous within a GPUs even when it is limited to running a single OS application at a time, because it helps simplify memory allocation and reduces memory fragmentation. After an entry is allocated in the PRT a memory request is forwarded to the memory management unit (MMU) ❽ for virtual to physical address translation and from there over a crossbar interconnect to the appropriate memory partition unit. As will be expanded upon in Section 4.3, the memory partition units contain a bank of L2 cache along with a memory access scheduler. Along with information about which physical memory address to access and how many bytes to read, the memory request contains a "subid" that can be used to lookup the entry in the PRT containing information about the request when the memory request returns to the core.

Once a memory request response for the load is returned to the core it is passed by the MMU to the fill unit ❾. The fill unit in turn uses the subid field in the memory request to lookup information about the request in the PRT. This includes information that can be passed by the fill unit to the load/store unit via the arbiter ❷ to reschedule the load which is then guaranteed to hit in the cache by locking the line in the cache after it has been placed into the data array ❺.

Cache Write Operations

The L1 data cache in Figure 4.1 can support both write through and write back policies. Thus, store instructions (writes) to global memory can be handled in several ways. The specific memory space written to determines whether the write is treated as write through or write back. Accesses to global memory in many GPGPU applications can be expected to have very poor temporal locality as commonly kernels are written in such a way that threads write out data to a large array right before exiting. For such accesses a write through with no write allocate [Hennessy and Patterson, 2011] policy might make sense. In contrast, local memory writes for spilling registers to the stack may show good temporal locality with subsequent loads justifying a write back with write allocate policy [Hennessy and Patterson, 2011].

The data to be written either to shared memory or global memory is first placed write data buffer (WDB) **10**. For uncoalesced accesses or when some threads are masked off, only a portion of a cache block is written to. If the block is present in the cache the data can be written to the data array via the data crossbar **6**. If the data is not present in the cache the block must first be read from the L2 cache or DRAM memory. Coalesced writes which completely fill a cache block may bypass the cache if they invalidate tags for any stale data in the cache.

Note that the cache organization described in Figure 4.1 does *not* support cache coherence. For example, suppose a thread executing on SM 1 reads memory location A and the value is stored in SM 1's L1 data cache and then another thread executing on SM 2 writes memory location A. If any thread on SM 1 subsequently reads memory location A before it is evicted from SM 1's L1 data cache it will obtain the old value instead of the new value. To avoid this issue, NVIDIA GPUs starting with Kepler only permitted local memory accesses for register spills and stack data or read-only global memory data to be placed in the L1 data cache. Recent research has explored how to enable coherent L1 data caches on GPUs [Ren and Lis, 2017, Singh et al., 2013] and the need for clearly defined GPU memory consistency models [Alglave et al., 2015].

4.1.2 L1 TEXTURE CACHE

Recent GPU architectures from NVIDIA combine the L1 data cache and texture cache to save area. To better understand how such a cache may work it is first necessary to understand a bit about the design of a stand alone texture cache. The details covered here should help provide additional intuition into how to develop microarchitectures for throughput processors. Much of the discussion here is based upon a paper by Igehy et al. [1998] that aimed to fill in for a lack of literature on how industrial texture cache designs tolerate long off-chip latencies for cache misses. Recent industry GPU patents [Minken et al., 2010, Minken and Rubinstein, 2003] describe closely related designs. As the focus of this book is not on graphics we provide only a brief summary of the texture operations that motivate the inclusion of texture caches.

In 3D graphics it is desirable to make scenes look as realistic as possible. To achieve this realism with the high frame rates required for real-time rendering, graphics APIs employ a technique called texture mapping [Catmull, 1974]. In texture mapping an image, called a texture, is applied to a surface in a 3D model to make the surface look more realistic. For example, a texture might be used to give the appearance of natural wood to a table in a scene. To implement texture mapping the rendering pipeline first determines the coordinates of one or more samples within the texture image. These samples are called texels. The coordinates are then used to find the address of the memory locations containing the texels. As adjacent pixels on the screen map to adjacent texels, and as it is common to average the values of nearby texels, there is significant locality in texture memory accesses that can be exploited by caches [Hakura and Gupta, 1997].

Figure 4.2 illustrates the microarchitecture of an L1 texture cache as described by Igehy et al. [1998]. In contrast to the L1 data cache described in Section 4.1.1, the tag array ❷ and data array ❺ are separated by a FIFO buffer ❸. The motivation for this FIFO is to hide the latency of miss requests that may need to be serviced from DRAM. In essence, the texture cache is designed assuming that cache misses will be frequent and that the working set size is relatively small. To keep the size of the tag and data arrays small, the tag array essentially runs ahead of the data array. The contents of the tag array reflect what the data array in the future after an amount of time roughly equal to the round trip time of a miss request to memory and back. While throughput is improved relative to regular CPU design with limited capacity and miss handling resources, both cache hits and misses experience roughly the same latency.

In detail, the texture cache illustrated in Figure 4.2 operates as follows. The load/store unit ❶ sends the computed addresses for texels to perform a lookup in the tag array ❷. If the access hits, a pointer to the location of the data in the data array is placed in an entry at the tail of the fragment FIFO ❸ along with any other information required to complete the texture operation. When the entry reaches the head of the fragment FIFO a controller ❹ uses the pointer to lookup the texel data from the data array ❺ and return it to the texture filter unit ❻. While not shown in detail, for operations such as bilinear and trilinear filtering (mipmap filtering) there are actually four or eight texel lookups per fragment (i.e., screen pixel). The texture filter unit combines the texels to produce a single color value which is returned to the instruction pipeline via the register file.

In the event of a cache miss during tag lookup, the tag array sends a memory request via the miss request FIFO ❽. The miss request FIFO sends requests to lower levels of the memory system ❾. DRAM bandwidth utilization in GPU memory systems can be improved by the use of memory access scheduling techniques [Eckert, 2008, 2015]. that may service memory requests out-of-order to reduce row switch penalties. To ensure the contents of the data array ❺ reflect the time-delayed state of the tag array ❷, data must be returned from the memory system in order. This is accomplished using a reorder buffer ❿.

4.1.3   UNIFIED TEXTURE AND DATA CACHE

In recent GPU architectures from NVIDIA and AMD caching of data and texture values is performed using a unified L1 cache structure. To accomplish this in this most straightforward way, only data values that can be guaranteed to read-only are cached in the L1. For data that follows this restriction the texture cache hardware can be used almost unmodified except for changes to the addressing logic. Such a design is described in a recent patent [Heinrich et al.,

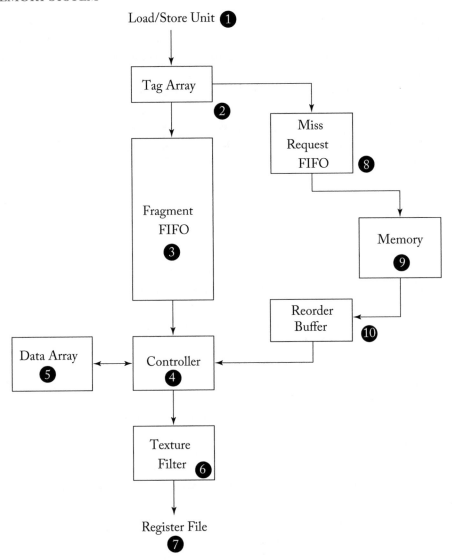

Figure 4.2: L1 texture cache (based in part on Figure 2 in [Igehy et al., 1998]).

2017]. In AMD's GCN GPU architecture all vector memory operations are processed through the texture cache [AMD, 2012].

4.2   ON-CHIP INTERCONNECTION NETWORK

To supply the large amount of memory bandwidth required to supply the SIMT cores, high-performance GPUs connect to multiple DRAM chips in parallel via memory partition units (described in Section 4.3). Memory traffic is distributed across the memory partition units using address interleaving. An NVIDIA patent describes address interleaving schemes for balancing traffic among up to 6 memory partitions at granularities of 256 bytes or 1,024 bytes [Edmondson and Van Dyke, 2011].

The SIMT cores connect to the memory partition units via an on-chip interconnection network. The on-chip interconnection networks described in recent patents for NVIDIA are crossbars [Glasco et al., 2013, Treichler et al., 2015]. GPUs from AMD have sometimes been described as using ring networks [Shrout, 2007].

4.3   MEMORY PARTITION UNIT

Below, we describe the microarchitecture of a memory partition unit corresponding to several recent NVIDIA patents. In terms of historical context, these patents were filed about a year prior to the release of NVIDIA's Fermi GPU architecture. As shown in Figure 4.3, each memory partition unit contains a portion of the second-level (L2) cache along with a one or more memory access schedulers also called a "frame buffer," or FB, and a raster operation (ROP) unit. The L2 cache contains both graphics and compute data. The memory access scheduler reorders memory read and write operations to reduce overheads of accessing DRAM. The ROP unit is primarily used in graphics operation such as alpha blending and supports compression of graphics surfaces. The ROP unit also supports atomic operations like those found in the CUDA programming model. All three units are tightly coupled and will be described below in some detail.

4.3.1   L2 CACHE

The L2 cache design includes several optimizations to improve overall throughput per unit area for the GPU. The L2 cache portion inside each memory partition is composed of two slices [Edmondson et al., 2013]. Each slice contains separate tag and data arrays and processes incoming requests in order [Roberts et al., 2012]. To match the DRAM atom size of 32 bytes in GDDR5, each cache line inside the slice has four 32-byte sectors. Cache lines are allocated for use either by store instructions or load instructions. To optimize throughput in the common case of co-alesced writes that completely overwrite each sector on a write miss no data is first read from memory. This is quite different from how CPU caches are commonly described in standard computer architecture textbooks. How uncoalesced writes, which do not completely cover a sector, are handled is not described in the patents we examined, but two solutions are storing byte-level valid bits and bypassing the L2 entirely. To reduce area of the memory access scheduler, data that is being written to memory is buffered in cache lines in the L2 while writes awaiting scheduling.

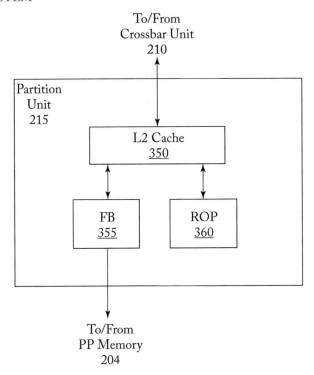

Figure 4.3: Memory partition unit overview (based on Figure 3B in Edmondson et al. [2013]).

4.3.2   ATOMIC OPERATIONS

As described in Glasco et al. [2012] the ROP unit includes function units for executing atomic and reduction operations. A sequence of atomic operations accessing the same memory location can be pipelined as the ROP unit includes a local ROP cache. Atomic operations can be used for implementing synchronization across threads running in different thread blocks.

4.3.3   MEMORY ACCESS SCHEDULER

To store large amounts of data GPUs employ special dynamic random access memory (DRAM) such as GDDR5 gdd. DRAM stores individual bits in small capacitors. To, for example, read values from these capacitors a row of bits, called a page, is first read into a small memory structure called a row buffer. To accomplish this operation the bitlines connecting the individual storage capacitors to the row buffer, and which have capacitance themselves, must first be precharged to a voltage half way between 0 and the supply voltage. When the capacitor is connected to the bit line through an access transistor during an activate operation the voltage of the bit line is pulled either up or down slightly as charge flow in or out of the storage cell from the bitline. A sense amplifier then amplifies this small change until a clean logic 0 or 1 is read. The process of reading the values into the row buffer refreshes the values stored in the capacitors. The precharge and activate operations introduce delays during which no data can be read or written to the DRAM array. To mitigate these overheads DRAMs contain multiple banks, each with their own row buffer. However, even with multiple DRAM banks it is often not possible to completely hide the latency of switching between rows when accessing data. This has led to the use of memory access schedulers [Rixner et al., 2000, Zuravleff and Robinson, 1997] that reorder DRAM memory access requests so as to reduce the number of times data must be moved between the row buffers and DRAM cells.

To enable access to DRAM, each memory partition in the GPU may contain multiple memory access schedulers [Keil and Edmondson, 2012] connecting the portion of L2 cache it contains to off-chip DRAM. The simplest approach for doing this would be for each slice of the L2 cache to have its own memory access scheduler. Each memory access scheduler contains separate logic for sorting read requests and write requests ("dirty data notifications") sent from the L2 cache [Keil et al., 2012]. To group together reads to the same row in a DRAM bank, two separate tables are employed. The first, called the read request sorter, is a set associative structure accessed by memory address and maps all read requests to the same row in a given bank to single pointer. The pointer is used to lookup a list of individual read requests in a second table called the read request store.

4.4 RESEARCH DIRECTIONS FOR GPU MEMORY SYSTEMS

4.4.1 MEMORY ACCESS SCHEDULING AND INTERCONNECTION NETWORK DESIGN

Yuan et al. [2009] explore memory access scheduler design for GPUs running GPU computing applications written in CUDA. They observe that requests generated by a single streaming multiprocessor (SM) have row-buffer locality. A sequence of memory requests to a given memory partition is said to have row buffer locality if requests that appear nearby in the sequence access the same DRAM row in the same DRAM bank. However, when memory requests from one SM are sent to the memory partitions they are intermixed with requests from other SM sending requests to the same memory partition. The result is that the row buffer locality of the sequence of requests entering the memory partition is lower. Yuan et al. [2009] propose reducing the complexity of memory access scheduling by modifying the interconnection network to maintain row buffer locality. They do this by introducing arbitration policies that prioritize granting packets containing memory requests from the same SM or with similar row-bank addresses.

Bakhoda et al. [2010, 2013] explore the design of on-chip interconnection networks for GPUs. This interconnect connects the streaming multiprocessors to the memory partitions. They argue that as the number of SMs increase it will become necessary to employ scalable topologies such as meshes. They explore how network-on-chip design impacts system throughput and find that throughput of many CUDA applications is relatively insensitive to interconnect latency. They analyze the interconnect traffic and find it has a many-to-few-to-many pattern. They propose a more restricted scalable topology consisting of "half-routers" that reduces the area cost of routers by leveraging this traffic pattern.

4.4.2   CACHING EFFECTIVENESS

Bakhoda et al. [2009] studied the impact of adding L1 and/or L2 caches for global memory accesses to a CUDA-enabled GPU simulated using their GPGPU-Sim simulator and showed that while some applications benefited others did not.

Subsequent work by Jia et al. [2012] characterized the effectiveness of caching by enabling or disabling caches on NVIDIA Fermi GPU hardware and found similar results. It is observed that applications that read data into the scratchpad shared memory via the L1 cache do not benefit from enabling the L1 cache. Even excluding such applications Jia et al. [2012] observe that cache hit rates alone are insufficient to predict whether caching will improve performance. They find that instead it is necessary to consider the impact of caching on request traffic to the L2 caches (e.g., memory partitions). On the Fermi GPU they study, the L1 cache is not sectored and as a consequence enabling caching can incur larger 128-byte off-chip memory accesses on a miss. On memory bandwidth-limited applications this additional off-chip memory traffic can result in a reduction in performance. Jia et al. [2012] introduce a taxonomy of three forms of locality: within-warp, within-block, and cross-instruction. Within warp locality occurs when memory read accesses from a single load executed by different threads within a single warp access the same cache block. Within block locality occurs when memory read accesses from a single load executed by threads in different warps from the same thread block access the same cache block. Cross-instruction locality occurs when memory read access from different load instructions execute by threads in the same thread block access the same cache block. Jia et al. [2012] introduce a compile time algorithm using this taxonomy to help infer when enabling caching is helpful for individual load instructions.

4.4.3   MEMORY REQUEST PRIORITIZATION AND CACHE BYPASSING

Following up on the above characterization study [Jia et al., 2012] and work by Rogers et al. [2012] which demonstrated warp scheduling can improve cache effectiveness (described in Section 5.1.2), Jia et al. [2014] proposed memory request prioritization and cache bypassing techniques for GPUs. Caches with low associativity relative to number of threads can suffer from significant conflict misses [Chen and Aamodt, 2009]. Jia et al. [2014] noted that several GPGPU applications written in CUDA include array indexing that causes an individual memory request from a single warp to result in misses that map to the same cache set when using a standard modulo cache set indexing function [Hennessy and Patterson, 2011, Patterson and Hennessy, 2013]. Jia et al. [2014] call this intra-warp contention. Assuming space in the cache is allocated when a miss is detected[1] and a limited number of miss-status holding registers[2] intra-warp contention can lead to memory pipeline stalls.[3] To address intra-warp contention Jia et al. [2014], propose bypassing the L1 data cache when a miss occurs and a cache block cannot be allocated due to an associativity stall. An associativity stall occurs when all blocks in a cache set are reserved to provide space for data that will be supplied by outstanding cache misses.

Jia et al. [2014] also examine what they call cross-warp contention. This form of cache contention results when one warp evicts data brought in by another warp. To address this form of contention, Jia et al. [2014] suggest employing a structure they call a "memory request prioritization buffer" (MRPB). Like CCWS [Rogers et al., 2012] the MRPB reduces capacity misses by modifying the order of accesses to the cache so as to increase locality. However, unlike CCWS which achieves this indirectly via thread scheduling, the MRPB attempts to increase locality by changing the order of individual memory accesses after threads have been scheduled.

The MRPB implements memory request reordering right before the first-level data cache. The input of the MRPB is feed memory requests generated in the instruction issue pipeline stage after memory request coalescing has been performed. The output of the MRPB feeds memory requests into the first-level cache. Internally, the MRPB contains several parallel first-in, first-out (FIFO) queues. A cache request is distributed to these FIFOs using a "signature." Among several options they evaluated they found the most effective signature was to use "warp ID" (a number between 0 to the maximum number of warps that can run on a streaming multiprocessor). The MRPB employs a "drain policy" to determine which FIFO to select a memory request from to use to access the cache next. Among several options explored the best version was a simple fixed-priority scheme in which each queue is assigned a static priority and the queue with highest priority that contains requests is served first.

A detailed evaluation shows the combined mechanism of bypassing and reordering using the MRPB achieves geometric mean speedup of 4% over a 64-way 16 KB. Jia et al. [2014] also perform some comparisons with CCWS showing larger improvements. We note in passing that the evaluation in Rogers et al. [2012] employed a baseline architecture with a more sophisticated set index hashing function[4] to reduce the impact of associativity stalls. Also, subsequent, work by Nugteren et al. [2014] worked to reverse engineer the details of the actual set index hash function employed in the NVIDIA Fermi architecture and found it uses XOR-ing (which also tends to reduce such conflicts).

Similar to Rogers et al. [2013], Jia et al. [2014] show that their programmer transparent approach to improving performance can narrow the gap between simple code that uses caches and more highly optimized code that uses the scratchpad shared memory.

---

[1]The default in GPGPU-Sim where it is used to avoid protocol deadlock.
[2]Consistent with a limited set of pending request table entries—see Section 4.1.1.
[3]GPGPU-Sim version 3.2.0, used by Jia et al. [2014], does not model instruction replay described in Sections 3.3.2 and 4.1.
[4]See `cache_config::set_index_hashed` in https://github.com/tgrogers/ccws-2012/blob/master/simulator/ccws_gpgpu-sim/distribution/src/gpgpu-sim/gpu-cache.cc

Arunkumar et al. [2016] explore the effect of bypassing and varying the cache line size, based on the level of memory divergence present in a static instruction. They use observed reuse distance patterns and memory divergence degree to predict bypassing and optimal cache line size.

Lee and Wu [2016] propose a control-loop based cache bypassing method that attempts to predict reuse behavior on an instruction-by-instruction basis at runtime. The reuse behavior of cache lines are monitored. If cache lines loaded by a particular program counter do not experience enough reuse, accesses for that instruction are bypassed.

4.4.4 EXPLOITING INTER-WARP HETEROGENEITY

Ausavarungnirun et al. [2015] propose a series of improvements at the GPU's shared L2 and memory controller that mitigate memory latency divergence in irregular GPU applications. The techniques, collectively named Memory Divergence Correction (MeDiC), exploit the observation that there is heterogeneity in the level of memory latency divergence across warps in the same kernel. Based on how they interact with the shared L2 cache, each warp in a kernel can be characterized as all/mostly hit, all/mostly miss, or balanced. The authors demonstrate that there is little benefit in having warps that are not all hit, since warps the mostly hit must wait for the slowest access to return before they are able to proceed. They also demonstrate that queueing latency at the L2 cache can have a non-trivial performance impact and that this effect can be mitigated by bypassing the L2 cache for all requests (even those that may hit) for all warps that are not all-hit. This decreases the access latency for all-hit warps by reducing queueing delay. In addition to the adaptive bypassing technique, they propose modifications to the cache replacement policy and the memory controller scheduler in an attempt to minimize latency for warps detected to be all-hit warps. They also demonstrate that even for warps that are all-hit, the difference in queueing delay among L2 cache banks can cause additional potentially avoidable queuing delay since there is an unbalance in the queuing delay between L2 banks.

The microarchitectural mechanism proposed by the authors consists of four components: (1) a warp-type detection block—which classifies warps in the GPU as being one of the five potential types: All-miss, mostly-miss, balanced, mostly-hit, or all-hit; (2) a warp-type-aware bypass logic block which decides if requests should bypass the L2 cache; (3) a warp-type-aware insertion policy, which determines where insertions in the L2 will be placed in the LRU stack; and (4) a warp-type-aware memory scheduler that orders how L2 misses/bypasses are sent to DRAM.

The detection mechanism operates by sampling the hit ratio of each warp (total hits/accesses) on an interval basis. Based on this ratio, the warp takes on one of the five classifications listed above. The exact hit ratios that determine the boundaries for these classifications are tuned dynamically for each workload. During the classification interval, no request bypasses the cache in order to react to phase changes in each warp's L2 characteristic.

The bypassing mechanism sits in front of the L2 cache and receives memory requests tagged with the warp-type that generated them. This mechanism attempts to eliminate accesses from all-miss warps and transforms mostly-miss warps into all-miss warps. The block simply sends all requests marked as coming from all-miss and mostly-miss warps directly to the memory scheduler.

The cache management policy of MeDiC operates by changing where requests returned from DRAM are placed in the L2's LRU stack. Cache lines requested by a mostly-miss warp are inserted into the LRU position, while all other requests are inserted into the traditional MRU position.

Finally, MeDic modifies the baseline memory request scheduler to contain two memory access queues: a high-priority queue for all-hit and mostly-hit warps and a low-priority queue for balanced, mostly-miss, and all-miss warps. The memory scheduler simply prioritizes all requests in the high-priority queue of any of the requests in the low priority queue.

4.4.5 COORDINATED CACHE BYPASSING

Xie et al. [2015] explore the potential to selectively enable cache bypassing for improving cache hit rates. They employ profiling to determine for each static load instruction in the GPGPU application whether it has good locality, poor locality, or moderate locality. They mark the instructions accordingly. Load operations marked as having good locality are permitted to use the L1 data cache. Load operations marked as having poor locality are always bypassed. Load instructions marked with moderate locality employ an adaptive mechanism that works as follows. The adaptive mechanism operates at thread block granularity. For a given thread block, all moderate locality loads executed are treated uniformly. They either use the L1 or bypass. The behavior is determined at the time the thread blocks are launched based upon a threshold that is adapted online using a performance metric that takes account of L1 cache hits and pipeline resource conflicts. Their evaluation shows this approach improves cache hit rates significantly more than static warp limiting.

4.4.6 ADAPTIVE CACHE MANAGEMENT

Chen et al. [2014b] propose coordinated cache bypassing and warp throttling that takes advantage of both warp throttling and cache bypassing to improve performance on highly cache-sensitive applications. The proposed mechanism detects cache contention and memory resource contention at runtime, and coordinates throttling and bypassing policy accordingly. The mechanism implements cache bypassing via an existing CPU cache bypassing technique of protection distance, which prevents a cache line from being evicted for a number of accesses. Upon insertion into the cache, the line is assigned a protection distance and counters track the remaining protection distance for lines. Once the remaining protection distance reaches 0, the line is no longer protected and can be evicted. When a new memory request attempts to insert a new line into a set with no unprotected lines, the memory request bypasses the cache.

The protection distance is set globally and the optimal value differs between workloads. In this work, Chen et al. [2014b] sweep the static protection distance and demonstrate that GPU workloads are relatively insensitive to the protection distance value.

4.4.7   CACHE PRIORITIZATION

Li et al. [2015] observe that warp throttling optimize L1 cache hit rate while potentially leaving other resources such as off-chip bandwidth and L2 cache significantly underutilized. They propose a mechanism of assigning tokens to warps to determine which warps can allocate lines into the L1 cache. Additional "non-polluting warps" are not given a token so that while they can execute they are not permitted to evict data from the L1. This leads to an optimization space where both the number of warps that can be scheduled (W) and the number that have tokens (T) can be set to less than the maximum number of warps that can execute. They show that statically selecting the optimal value of W and T enables a 17% improvement over CCWS with static warp limiting.

Based on this observation, Li et al. [2015] explore two mechanisms to learn the best values for W and T. The first approach is based upon the idea of maintaining high thread level parallelism while increasing cache hit rates. In this approach, called dynPCALMTLP, a sampling period runs a kernel with W set to the maximum number of warps and then varies T across different SIMT cores. The value of T that achieves the maximum performance is then selected. This leads to comparable performance to CCWS with significantly less area overhead. The second approach, called dynPCALCCWS, initially uses CCWS to set W then uses dynPCALMTLP to determine T. Then it monitors resource usage of shared structures to dynamically increase or decrease W. This leads to an 11% performance gain versus CCWS.

4.4.8   VIRTUAL MEMORY PAGE PLACEMENT

Agarwal et al. [2015] consider the implications of supporting cache coherence across multiple physical memory types in a heterogeneous system including both capacity-optimized and bandwidth-optimized memory. Since DRAM optimized for bandwidth is more expensive in cost and energy than DRAM optimized for capacity, future systems are likely to include both. Agarwal et al. [2015] observe current OS page placement policies such as those deployed in Linux do not account for the non-uniformity of memory bandwidth. They study a future system in which a GPU can access low bandwidth/high capacity CPU memory at low latency—a penalty of 100 core cycles. Their experiments use a modified version of GPGPU-Sim 3.2.2 configured with additional MSHR resources to model more recent GPUs.

With this setup, they first find that for memory bandwidth limited applications there is significant opportunity to gain performance by using both CPU and GPU memory to increase aggregate memory bandwidth. They find less memory latency limited GPGPU applications for which this is not the case. Under the assumption that pages are accessed uniformly and when memory capacity of bandwidth-optimized memory is not a limitation, they show that allocating pages to memory regions in proportion to the regions' available memory bandwidth is optimal. Assuming capacity of bandwidth limited memory is not an issue, they find a simple policy of randomly allocating pages to bandwidth- or capacity-optimized memory with probability in proportion to memory bandwidth works in practice with real GPGPU programs. However, when bandwidth-optimized memory capacity is insufficient to meet application demands, they find it is necessary to refine the page placement to consider frequency of access.

To refine the page placement, they propose a system involving a profiling pass implemented using a modified version of the NVIDIA developer tools nvcc and ptxas along with an extension of the existing CUDA API to include page placement hints. Using profile-guided page placement hints obtains about 90% of the benefits of an oracle page placement algorithm. They leave page migration strategies to future work.

4.4.9   DATA PLACEMENT

Chen et al. [2014a] propose PORPLE, a portable data placement strategy that consists of specification language, a source-to-source compiler and an adaptive runtime data placer. They capitalize on the observation that with all the various flavors of memory available on the GPU, choosing what data should be placed where is difficult for the programmer to determine and is often not portable from one GPU architecture to the next. The goal of PORPLE is to be extensible, input-adaptive, and generally applicable to both regular and irregular data accesses. Their approach relies on three solutions.

The first solution is a memory specification language to help with extensibility and portability. The memory specification language describes all the various forms of memory on the GPU based on the conditions under which accesses to these spaces are serialized. For example, accesses to adjacent global data are coalesced, hence accessed concurrently, but accesses to the same bank of shared memory must be serialized.

The second solution is a source-to-source compiler named PORPLE-C which transforms the original GPU program into a placement agnostic version. The compiler inserts guards around accesses to memory, selecting the access that corresponds to the predicted best placement of the data.

Finally, to predict which data placement would be most optimal, they use PORPLE-C to find static access patterns through code analysis. When the static analysis cannot make a determination on the access pattern, the compiler generates a function that traces the runtime access patterns and attempts to make a prediction. This function is run on the CPU for a short period of time and helps determine the best GPU-based data placement prior to launching the kernel. In the scope of this work, the system only handles the placement of arrays, as they are the most prevalent data structure used in GPU kernels.

The lightweight model used to make the placement prediction in PORPLE generates an estimate of the number of transactions generated based on the serialization conditions of the memory. For memories that have a cache hierarchy, it uses a reuse distance estimation of cache hit rate. When multiple arrays share a cache, the estimate of how much cache is devoted to each array is based on a linear partitioning of the cache based on the size of the array.

4.4.10   MULTI-CHIP-MODULE GPUS

Arunkumar et al. [2017] note that the slowing of Moore's Law will result in slowing increases in GPU performance. They propose to extend performance scaling by building large GPUs out of smaller GPU modules on a multichip module (see Figure 4.4). They demonstrate it is possible to attain with 10% of the performance of a single large (and unimplementable) monolithic GPU by combining local caching of remote data, CTA scheduling to modules that considers locality and first-touch page allocation. According to their analysis this is 45% better performance than possible using the largest implementable monolithich GPU in the same process technology.

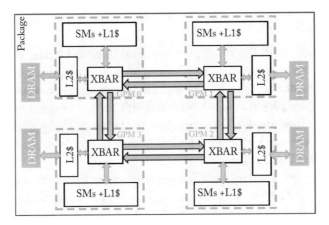

Figure 4.4: A multi-chip-module GPU (based on Figure 3 from Arunkumar et al. [2017]).

CHAPTER 5

Crosscutting Research on GPU Computing Architectures

This chapter details several research directions in GPGPU architecture that do not fit neatly into earlier chapters which focused on specific portions of the GPU architecture. Section 5.1 explores work on how threads are scheduled in GPUs. Section 5.2 looks at alternative programming methodologies and Section 5.4 examines work on heterogeneous CPU/GPU computing.

5.1 THREAD SCHEDULING

Contemporary GPUs are fundamentally different from CPUs in that they rely on massive parallelism. Independent of how a program is specified (e.g., using OpenCL, CUDA, OpenACC, etc.), workloads without extensive software-defined parallelism are not suitable for GPU acceleration. GPUs employ several mechanisms to aggregate and schedule all these threads. There are three primary ways threads on the GPU are organized and scheduled.

Assignment of Threads to Warps Since GPUs use SIMD units to execute threads defined by a MIMD programming model, the threads must be fused together for lock-step execution in the form of warps. In the baseline GPU architecture studied throughout this book, threads with consecutive thread IDs are statically fused together to form warps. Section 3.4.1 summarizes the research proposals on alternative thread arrangements within warps for better warp compaction.

Dynamic Assignment of Threadblocks to Cores Unlike in CPUs where threads can be assigned to hardware threads one at a time, in GPUs, work is assigned to GPU cores in bulk. This work-unit consists of multiple warps in the form of a threadblock. In our baseline GPU, threadblocks are assigned to cores in round-robin order. The core's resources (like warp-slots, register file, and shared memory space) are subscribed at the threadblock granularity. Due to the large amount of states associated with each threadblock, contemporary GPUs do not preempt their execution. Threads in the threadblock run to completion before their resources can be assigned to another threadblock.

Cycle-by-cycle Scheduling Decisions After a threadblock have been assigned to a GPU core, a collection of fine-grained hardware schedulers decides, at each cycle, which set of warps to fetch instructions, which warps to issue instructions for execution, and when to read/write operands for each issued instruction.

Scheduling Multiple Kernels Threadblock-level and cycle-by-cycle scheduling decisions can take place both within a kernel and across different kernels running concurrently on the same GPU. Legacy kernel scheduling limited just one kernel to be active on a GPU at a time. However, the introduction of NVIDIA's Streams and HyperQ scheduling mechanisms make the running of concurrent kernels possible. This situation is similar in some ways to multiprogramming on CPUs.

5.1.1 RESEARCH ON ASSIGNMENT OF THREADBLOCKS TO CORES

When a kernel is launched, threads within each kernel launch are grouped into threadblocks. A GPU-wide threadblock scheduling mechanism assigns each threadblock to one of the SIMT cores, based on resource availability. Each core has a fixed amount of scratchpad memory (called shared memory in CUDA or local memory in OpenCL), number of registers, slots for warps, and slots for threadblocks. At kernel launch time, all of these parameters are known for each threadblock. The most obvious threadblock scheduling algorithm is to assign threadblocks to cores in a round-robin fashion to maximize the number of cores involved. Threadblocks are continuously scheduled until at least one resource in each core is exhausted. Note that a kernel may be composed of more threadblocks than can be run on the GPU at once. As a result, some threadblocks in a kernel may not even be running on the GPU while others execute. Several research techniques have looked at trade-offs in the threadblock scheduling space.

Throttling at the Threadblock Level. Kayiran et al. [2013] propose throttling the number of threadblocks allocated to each core in order to reduce contention in the memory system caused by thread over-subscription. They develop an algorithm that monitors core idle cycles and memory delay cycles. The algorithm starts by assigning each core only half of its maximum threadblocks. Then the idle and memory delay cycles are monitored. If a core is predominantly waiting for memory, then no more threadblocks are assigned and existing threadblocks are potentially paused to stop them from issuing instructions. The technique achieves a course-grained parallelism throttling mechanism, which limits memory system interference and improves overall application performance, even though less CTAs are concurrently active.

Dynamically Tuning GPU Resources. Sethia and Mahlke [2014] propose Equalizer, a hardware runtime system that dynamically monitors resource contention and scales the number of threads, core frequency, and memory frequency to improve both energy consumption and performance. The system bases its decision on four parameters: (1) the number of active warps in an SM; (2) the number of warps waiting for data from memory; (3) number of warps ready to execute an arithmetic instruction; and (4) number of warps ready to execute a memory instruction. With these parameters, it first decides the number of warps to keep active on an SM, then based on this value and the values of the other three counters (which act as proxies for memory contention, compute intensity, and memory intensity) it decides how best to scale the frequency of both the core and the memory system.

Equalizer has two modes of operation: energy-saving mode and performance-enhancing mode. In energy-savings mode it saves energy by scaling back the under-utilized resource to minimize energy consumption while mitigating its impact on performance. In performance-enhancing mode, Equalizer boosts the bottleneck resource increasing performance in an energy-efficient manner.

They characterize a set of workloads from Rodinia and Parboil as being either compute intensive, memory intensive, cache sensitive, or unsaturated by examining the performance and energy tradeoffs associated with changing the memory frequency, compute frequency, and the number of the threads concurrently running. If minimizing energy (without sacrificing performance) is the goal, then compute intensive kernels should operate with a lower memory frequency and memory kernels should operate with a lower SIMT core frequency. This help reduce energy unnecessarily spent in the system that is not being fully utilized at the baseline rate.

Equalizer makes decisions about frequency and concurrency on an interval basis. The technique adds monitor hardware to each SIMT core that makes a local decision based on the four counters listed previously. It decides locally in each SIMT core what the three output parameters (number of CTAs, memory frequency, and compute frequency) should be for this epoch. It informs a global work distribution engine about the number of CTAs this SM should use, issuing new blocks if the SIMT core wants more work. If the SM should run with fewer CTAs, it pauses some of the CTAs on the core. After deciding the number of CTAs to run, each SIMT core submits a memory/compute voltage target to a global frequency manager which sets the chip-wide frequencies based on a majority function.

The local decision is made by observing the number of warps that are waiting to execute a memory instruction and the number of warps that are waiting to execute an ALU instruction. If the number of warps trying to wait for memory is greater than the number of warps in a CTA, then the number of CTAs running on this SIMT core is reduced, potentially helping the performance of cache-sensitive workloads. The SIMT core is considered memory (or compute) intensive if the number of warps ready to issue memory (or ALU) is more than the number of warps in a CTA. If there are fewer warps waiting on memory (or compute) than there are warps in a CTA, the SIMT core can still be considered ALU or memory bound if more than half of the active warps are waiting and there are not more than two warps waiting on memory. If this is the case, then the number of active CTAs on the core is incremented by one and the SIMT core is determined to be compute or memory bound based on if there are more compute waiting warps or more memory waiting warps.

Once the SIMT cores have made their local decisions, the frequencies of the memory and the cores are scaled by $\pm 15\%$ based on the mode of operation Equalizer is operating in.

5.1.2   RESEARCH ON CYCLE-BY-CYCLE SCHEDULING DECISIONS

Early Characterizations of Cycle-by-Cycle Scheduling.   Lakshminarayana and Kim [2010] explore numerous warp-scheduling policies in the context of an early GPU without hardware managed caches and show that, for applications that execute symmetric (balanced) dynamic instruction counts per warp, a fairness-based warp and DRAM access scheduling policy improves performance. This policy works well on the regular GPU workloads used in they study because regular memory requests between warps are both merged within the core and better exploit DRAM row-buffer locality. The paper also characterizes several other warp scheduling policies, including ICOUNT, which was first proposed by Tullsen et al. [1996] for simultaneously multi-threaded CPUs. ICOUNT is designed to improve system throughput by prioritizing the fastest progressing warp (or thread). Lakshminarayana and Kim [2010] show that prioritizing only a few warps in their early cache-less GPU on early, regular workloads generally does not improve performance.

Two-Level Scheduling.   Gebhart et al. [2011c] introduce the use of a two-level scheduler to improve energy efficiency. Their two-level scheduler divides warps in a core into two pools: an active pool of warps that are considered for scheduling in the next cycle and an inactive pool of warps that are not. A warp transition out of the active pool whenever it encounters a compiler identified global or texture memory dependency and back into the active pool in round-robin fashion from the inactive pool. Selecting from a smaller pool of warps every cycle reduces the size and energy consumption of the warp selection logic.

The two-level scheduler proposed by Narasiman et al. [2011] focuses on improving performance by allowing groups of threads to reach the same long latency operation at different times. This helps ensure cache and row-buffer locality within a fetch group is maintained. The system can then hide long latency operations by switching between fetch groups. In contrast, Cache-Conscious Warp Scheduling (see below) focuses on improving performance by adaptively limiting the amount of multithreading the system can maintain based on how much intra-warp locality is being lost.

Cache-Conscious Warp Scheduling.   Rogers et al. [2012] categorize the memory locality that exists in GPU kernels as being *intra-warp*, where a warp load then references its own data, or *inter-warp*, where a warp shares data with other warps. They demonstrate that intra-warp locality is most common form of locality that occurs in cache-sensitive workloads. Based on this observation, they proposed a cache-conscious wavefront scheduling (CCWS) mechanism to exploit this locality by throttling the number of warps actively scheduled based on memory system feedback.

Actively scheduling between fewer warps enables each individual warp to consume more cache space and reduces L1 data cache contention. In particular, throttling occurs when workloads with locality are thrashing the cache. To detect this thrashing, CCWS introduces a lost-locality detection mechanism which is based on replacement victim tags from the L1 data cache.

Figure 5.1 plots the high-level microarchitecture of CCWS. On every eviction from the cache, the victim's tag is written to warp-private victim tag array. Each warp has its own victim tag array, because CCWS is only concerned with detecting intra-warp locality. On every subsequent cache miss, the victim tag array for the missing warp is probed. If the tag is found in the victim tags, then some intra-warp locality has been lost. CCWS makes the assumption that this warp might have been able to hit on this line if the warp had more exclusive access to the L1 data cache, and therefore could benefit potentially benefit from throttling.

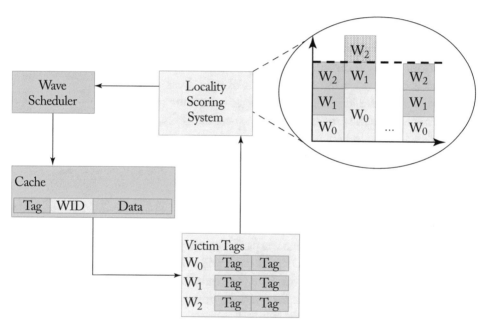

Figure 5.1: Cache-conscious wavefront scheduling microarchitecture.

To reflect this loss in locality, a signal is sent to the scheduling system. The issue scheduler uses a locality scoring system to approximate how much locality each warp in the system has lost, which is an approximation of how much additional cache capacity each warp requires. All warps in the locality scoring system are assigned an initial score, with the assumption that all warps require the same cache capacity and that no throttling occurs (the stacked bars in Figure 5.1). As time passes and lost locality is detected, the scores for individual warps are increased. In the example in Figure 5.1, warp 0 has experienced a loss in locality and its score has been increased. The increase in its score has pushed warp 3 over a threshold, which will prevent it from issuing L1 data cache requests, effectively throttling the number of warps actively scheduled on the core. Over time, if no locality is lost then the score for warp 0 is decreased until warp 2 is able to fall below the threshold and is able to issue memory requests again.

CCWS goes on to demonstrate the sensitivity cache hit rate has to scheduling decisions by comparing various scheduling mechanisms to cache replacement policies. The paper demonstrates that the decision space available to the warp scheduler is much greater than the relatively constrained decision space made by a replacement policy. The paper goes on to further demonstrate that the CCWS scheme, using an LRU replacement policy can improve cache hit rate more than prior scheduling mechanisms, even if they use a Belady optimal cache-replacement policy.

Rogers et al. [2013] propose Divergence-Aware Warp Scheduling (DAWS), which extends CCWS with a more accurate estimate of each warp's cache footprint. DAWS capitalizes on the fact that most intra-warp locality on GPU workloads occurs in loops. DAWS creates a per-warp cache footprint estimate for warps in loops. DAWS pre-emptively throttles the number of warps in loops based on the per-warp predicted loop footprint. DAWS also adapts its cache-footprint estimate with the level of control flow divergence experienced by each warp. Threads in a warp that have left the loop no longer contribute to the footprint estimate. DAWS goes on to further explore the programmability aspect of GPUs, demonstrating that with a more intelligent warp scheduler, a benchmark with no optimization on memory transfer (for example, using shared memory instead of cache) can come very close to matching a GPU-optimized version of the same benchmark.

Prefetching-Aware Warp Scheduling.   Jog et al. [2013b] explore a prefetching-aware warp scheduler on the GPU. They base the scheduler on the two-level scheduling mechanism, but form fetch groups from non-consecutive warps. This policy increases the amount of bank-level parallelism in the DRAM since one DRAM bank will not be queried for contiguous accesses by the prefetcher. They further extend this idea to manipulate the prefetcher based on the warp-group assignment. By prefetching data for warps in other groups they can improve row-buffer locality and provide spacing between the prefetch request and the demand for data.

CTA-Aware Scheduling.   Jog et al. [2013a] propose a CTA-aware warp scheduler that also builds on the two-level scheduler to form fetch groups based on the selectively combining CTAs. They exploit several CTA-based properties to improve performance. They employ a throttling prioritization technique that throttles the number of active warps in the core, similar to the other throttling schedulers. Combined with throttling, they exploit inter-CTA page locality among CTAs on different cores. Under a locality-aware only CTA scheduler, consecutive CTAs will often access the same DRAM bank at the same time, decreasing bank-level parallelism. They combine this with a prefetching mechanism to also improve DRAM row locality.

Impact of Scheduling on Branch Divergence Mitigation Techniques.   Meng et al. [2010] introduce Dynamic Warp Subdivision (DWS) which splits warps when some lanes hit in cache and some lanes do not. This scheme allows individual scalar threads that hit in cache to make progress even if some of their warp peers miss. DWS improves performance by allowing run-ahead threads to initiate their misses earlier and creates a pre-fetching effect for those left behind. DWS attempts to improve intra-warp locality by increasing the rate data is loaded into the cache.

Fung et al. [2007] explore the impact of warp scheduling policy on the effectiveness of their Dynamic Warp Formation (DWF) technique. DWF attempts to mitigate control flow divergence by dynamically creating new warps when scalar threads in the same warp take different paths on a branch instruction. They propose five schedulers and evaluate their effect on DWF.

Fung and Aamodt [2011] also propose three thread block prioritization mechanisms to compliment their Thread Block Compaction (TBC) technique. The prioritization mechanisms attempt to schedule threads within the same CTA together. Their approach is similar to concurrent work on two-level scheduling proposed by Narasiman et al. [2011], except thread blocks are scheduled together instead of fetch groups.

Section 3.4 contains more detailed summaries for DWS, DWF, and TBC.

Scheduling and Cache Re-Execution.   Sethia et al. [2015] introduce Mascar which attempts to better overlap computation with memory accesses in memory intensive workloads. Mascar consists of two intertwined mechanisms.

- A memory aware warp scheduler (MAS) that prioritizes the execution of a single warp when MSHR and L1 miss queue entries in the core are oversubscribed. This prioritization helps improve performance even when workloads do not contain data locality by enabling warps executing on the in-order core to reach their computation operations faster, enabling overlap of the prioritized warp's computation with other warp's memory accesses.

- A cache access re-execution (CAR) mechanism that helps avoiding L1 data cache thrashing by enabling L1 data cache hits-under-misses when warps with data in the cache are blocked from issuing because low-locality accesses are stalling the memory pipeline.

MAS has two modes of operation: equal priority (EP) and memory access priority (MAP) mode. The system switches between EP and MP based on how full the L1 MSHRs and memory miss queue is. Once these structures are almost full, the system switches to MP mode. MAS contains two queues, one for memory warps (warps attempting to issue a memory instruction) and one for compute warps (warps attempting to issue other types of instructions). Within each queue, warps are scheduled in greedy-then-oldest order. The tracking of memory dependent instructions is done by augmenting the scoreboard to indicate when an output register is filled based on a load. The scheduler operates in EP mode when it is observed that the workload is balanced and the memory system is not oversubscribed. In EP mode, the scheduling mechanism prioritizes memory warps first. Since the memory system is not oversubscribed, it is predicted that initiating memory accesses early will improve performance. When operating in MAP mode, the scheduler prioritizes compute warps to better overlap available compute with the bottlenecked memory system. Only one of the memory warps, the "owner warp" is allowed to issue memory instructions until it reaches an operation that is dependent on a pending memory request.

In addition to the scheduling mechanism, Sethia et al. [2015] show that memory intensive kernels perform at a much lower fraction of their peak IPC than compute intensive kernels. They illustrate that in the memory intensive applications, a large fraction of cycles are spent with the SIMT core's load store unit stalled due to memory back-pressure caused from excessive memory accesses. While the LSU is stalled, there is a significant fraction of time where data for ready warps is in the L1 data cache, but the warps are unable to issue because the LSU is backed up with memory requests other warps. The cache access re-execution (CAR) mechanism seeks to remedy this behavior by providing a buffer on the side of the LSU pipeline which stores stalled memory instructions and allows others issue into the LSU. Requests are processed from the re-execution queue only if the LSU is not stalled and has no new requests to issue unless the re-execution queue is full, in which case accesses in the re-execution queue are prioritized until space frees up in the queue.

When the re-execution queue is combined with the memory aware scheduler, special care needs to be taken since requests in the re-execution queue can come from warps other than the prioritized owner warp. When operating in MAP mode, non-owner warp's requests sent from the re-execution queue to the L1 are further delayed when they do not hit in the L1. In particular, when a request from a non-owner warp misses in the L1, the request is not relayed to the L2 cache, but rather is reinserted into the tail of the re-execution queue.

5.1.3   RESEARCH ON SCHEDULING MULTIPLE KERNELS

Supporting Pre-emption on GPUs.   Park et al. [2015] tackles the challenge of supporting preemptive multitasking on GPUs. It employs a more relaxed definition of idempotence to enable flushing of computation within a threadblock. The more relaxed definition of impotence involves detecting whether execution has been idempotent from the start of thread execution. Their proposal, Chimera, dynamically selects among three approaches to achieving a context switch for each threadblock:

- a full context save/store;

- waiting until the threadblock finishes; and

- simply stopping the threadblock without saving any context if, due to idempotence, the thread block can be restarted from the beginning safely.

Each context-switching technique provides a different tradeoff between the latency to switch and the impact on system throughput. To implement Chimera an algorithm estimates the subset of threadblocks that are currently running which can be stopped with minimum impact on system throughput while meeting a user specified context switch latency target.

5.1.4   FINE-GRAIN SYNCHRONIZATION AWARE SCHEDULING

ElTantawy and Aamodt [2018] explore the impact of warp scheduling when running codes that involve fine-grained synchronization. Using real GPU hardware they demonstrate that a significant overheads occur when threads spin waiting for locks. They point out that naively backing off execution of warps containing threads that have failed to acquire a lock can prevent or slow progress of other threads in the same warp already holding a lock. They propose a hardware structure for dynamically identifying which loops are involved in spin locks, which is made more challenging by the use of stack-based reconvergence [ElTantawy and Aamodt, 2016]. This structure uses a path history containing least significant bits of the program counter and a separate history of predicate register updates to accurately detect loops spinning on a lock. To reduce contention and improve performance they propose reducing the priority of warps identified as executing a spin loop when they execute the backwards branch of a spin loop after any threads in the warp that held locks have released those locks. They find this improves performance and reduces energy by 1.5× and 1.6×, respectively, vs. Lee and Wu [2014].

5.2   ALTERNATIVE WAYS OF EXPRESSING PARALLELISM

Fine-Grained Work Queues.   Kim and Batten [2014] propose the addition of a fine-grain hardware worklist to each SIMT core in the GPU. They exploit the observation that irregular GPGPU programs often perform best when implemented in software using a data-driven approach, where work is dynamically generated and balanced between threads, as opposed to a topological approach, where a fixed number of threads are launched—many of these threads do no useful work. The data-driven approach has the potential to increase work efficiency and load balancing but may suffer from poor performance without extensive software optimizations. This paper proposes an on-chip hardware worklist with support for load balancing both within the core and among cores. They use a thread waiting mechanism and rebalance tasks generated by threads on an interval basis. They evaluate their hardware mechanism on various implementations of the irregular applications in the lonestar GPU benchmark suite that make use of both topological and data-driven work distribution.

The on-core hardware worklist addresses two major problems with data-driven software worklists: (1) contention in the memory system when threads push out generated work and (2) poor load balancing as a result of statically partitioning work based on thread ID. Software implementations that do not rely on static partitioning suffer from memory contention on both pushes and pulls. Statically partitioning the work solves the pulling contention. The hardware worklist is distributed across multiple structures, reducing contention. It improves load balancing by dynamically redistributing generated work to threads before they become idle. The authors add special instructions to the ISA for pushing and pulling from the hardware queues. Each lane in the core is assigned a small, single ported SRAM that is used as the storage for work IDs used and generated by a given lane.

The paper proposes both an interval based and demand-driven (only redistributed on push/pull requests) work redistribution method and evaluates the former in depth. The interval based method redistributes the work on either a simple threshold basis or a more complex sorting basis. The threshold method classifies lanes with more work than the threshold as greedy (having too much work) and those with less work than the threshold as needy (not having enough work). A sorting pass then redistributes work from the greedy banks to the needy banks. The sorting-based technique is more complex, but achieves better load balancing since all needy banks can also donate work to other needy banks. Their technique also includes a global sorting mechanism that can be used to distribute work between cores. Additionally, the architecture provides support for virtualizing the hardware worklist, making it extendible to workloads that generate more dynamic work than the capacity available in the hardware structure.

Nested Parallel-pattern Based Programming.   Lee et al. [2014a] propose a Locality-Aware Mapping of Nested Parallel Patterns on GPUs which capitalized on the observation that there is not a universally optimal mapping of nested parallel computation to GPU threads. Algorithms with nested parallelism (such as a map/reduce operation) can have their parallelism exposed to the GPU at different levels, depending on how the GPU program is written. The authors exploit three generalizations of nested parallel mapping on GPUs:

- a 1D mapping, which parallelizes the outer loop of a sequential program;

- a thread block/thread mapping, which assigns each iteration of the sequential program's outer loop to a thread block and parallelizes the inner pattern across a thread block; and

- a warp-based mapping, which assigns each iteration of the outer loop to a warp and parallelizes the inner pattern across the warp.

This work proposes an automatic compilation framework which generates predicted performance scores based on locality and the degree of parallelism exposed in nested patterns to choose which mapping is best suited for a set of common nested parallel patterns. These patterns consist of collection operations such as map, reduce, foreach, filter, etc. The framework attempts to map threads to operations on each element of the collection. The framework handles the nesting of patterns by first assigning each nest level in the application to a dimension (x, y, z, etc.). A doubly nested pattern (i.e., a map containing a reduce) has two dimensions. The mapping then determines the number of threads in a given dimension in a CUDA thread block. After setting the dimensions and size of the thread block, the framework further controls the degree of parallelism in the kernel by assigning multiple elements to each thread using the notion of thread spanning and splitting. In a two-dimensional kernel (i.e., two levels of pattern nesting), if each dimension is assigned span(1), then each thread launched in the kernel is responsible for operating on only one element of the collection. This mapping exposes the greatest degree of parallelism. In contrast, span(all) indicates that each thread operates on all the elements in the collection. The span can be any number between (1) and (all). Span(all) is used in two special circumstances: when the size of a dimension is not known until after the kernel is launched (for example when the number of elements operated on in an inner pattern is determined dynamically) and when the pattern requires synchronization (for example, the reduce operation).

Since span(all) can severely limit the exposed parallelism and cause the GPU to be underutilized, the framework also provides the notion of a *split*. The split(2) indicates that each thread operates on half of the elements in a given dimension (think of it as span(all)/2). When split is used, the framework launches a second kernel (called a combiner kernel) to aggregate the results across the splits, producing the same result as if the kernel had been partitioned using span(all).

To select both the block size in each dimension and the split/span of each dimension, the framework uses a scoring algorithm based on hard and soft constraints. The algorithm sweeps the entire entire search space all possible dimensions, block sizes and spans. The search space is exponential to the level of loop nests. However, the base of the exponential is less than 100 and a typical kernel contains less than 3 levels. Thus, the space is completely searchable in just a few seconds. The search prunes configurations that violate hard constraints—i.e., those that cause incorrect execution, such as the maximum number of threads in a block being too high. It assigns weighted scores to soft constraints, such as ensuring that patterns sequential memory accesses are assigned to the x dimension to improve memory coalescing.

The framework also performs two common GPU optimizations: pre-allocating memory instead of dynamically allocating global memory in nested kernels and making use of shared memory when it determines that prefetching data into shared memory is beneficial for a nested pattern. Results demonstrate that the automatically generated code has competitive performance with expertly tuned code.

Dynamic Parallelism.   Wang and Yalamanchili [2014] characterize the overheads of using CUDA Dynamic Parallelism on Kepler GPU hardware and find these overheads can be substantial. Specifically, they identify several key issues limiting efficiency in the workloads they studied. First, the applications employed a very large number of device launched kernels. Second, each kernel typically had only 40 threads (little more than one warp). Third, while the code executed in each dynamic kernel is similar, the launch configuration differs resulting in a large amount of storage overhead for kernel configuration information. Finally, fourth, to enable concurrency the device launched kernels are placed in separate streams to exploit the 32 parallel hardware queues (Hyper-Q) supported on Kepler. They find these factors combine to yield very poor utilization.

Wang et al. [2016a] subsequently propose dynamic thread block launch (DTBL), which modifies the CUDA programming model to enable device launched kernels to share hardware queue resources enabling greater parallelism and better utilization of the GPU hardware. A key to their proposal is enabling dynamically launched kernels to be aggregated together with an existing kernel running the same code. This is supported by maintaining a linked list of aggregated thread blocks that are used by the modified hardware when launching kernels. They

evaluate DTBL by modifying GPGPU-Sim and find DTBL improves performance by 1.4×
versus CDP and 1.2× over highly tuned CUDA versions that do not employ CDP.

Wang et al. [2016b] then explore the impact of which SM dynamically launched thread
blocks are scheduled to. They find that by encouraging child thread blocks to be scheduled on
the same SM as parent SMs while considering workload distribution across SMs they were able
to improve performance by 27% vs. a naive round-robin distribution mechanism.

5.3 SUPPORT FOR TRANSACTIONAL MEMORY

This section summarizes the various proposals to support a transactional memory (TM) [Harris
et al., 2010, Herlihy and Moss, 1993] programming model on GPU architectures.

These proposals were motivated by the potential of a TM programming model to ease
the challenge of managing irregular, fine-grained communications between threads in GPU
applications with ample irregular parallelism. On modern GPUs, the application developers
may either coarsen the synchronization between threads via barriers, or they may attempt to use
single-word atomic operations, available on many modern GPUs, to implement fine-grained
locks for these communications. The former approach may involve significant changes to the
underlying algorithms, while the latter approach involves the uncertainty in development effort
with fine-grained locking, too risky for practical, market-driven software development (with
several exceptions). Enabling TM on GPUs simplifies synchronization, and provides a powerful
programming model that promotes fine-grained communication and strong scaling of parallel
workloads. This promise from TM hopes to encourage software developers to explore GPU
acceleration with these irregular applications.

Distinct Challenges for Supporting TM on GPUs.   The heavily multithreaded nature of
GPU introduces a new set of challenges to TM system designs. Instead of running tens of concurrent transactions with relatively large footprint—the focus of much recent research on TM
for multicore processors—TM system on a GPU aims to scale to tens of thousands of small
concurrent transactions. This reflects the heavily multithreaded nature of GPU, with tens of
thousands of threads working in collaboration, each performing a small task towards a common
goal. These small transactions are tracked at word-level granularity, enabling finer resolution of
conflict detection than cache blocks. Moreover, each per-core private cache in a GPU is shared
by hundreds of GPU threads. This drastically reduces the benefit of leveraging a cache coherence protocol to detect conflicts, a technique employed on most hardware transactional memory
designed for traditional CMPs with large CPU cores.

5.3.1 KILO TM

Kilo TM [Fung et al., 2011] is the first published hardware TM proposal for GPU architectures.

Kilo TM employs value-based conflict detection [Dalessandro et al., 2010, Olszewski
et al., 2007] to eliminate the need for global metadata for conflict detection. Each transaction

simply reads the existing data in global memory for validation—to determine if it has a conflict with another committed transaction. This form of validation leverages the highly parallel nature of the GPU memory subsystem, avoids any direct interaction between conflicting transactions, and detects conflicts at the finest granularity.

However, a native implementation of value-based conflict detection requires transactions to commit serially. To boost commit parallelism, Kilo TM incorporates ideas from existing TM systems [Chafi et al., 2007, Spear et al., 2008] and extended them with innovative solutions. In particular, Fung et al. [2011] introduced the *recency bloom filter*, a novel data structure that uses the notion of time and order to compress a large number of small item sets. Kilo TM uses this structure to compress the write-sets of all committing transactions. Each committing transaction queries the recency bloom filter for an approximate set of conflicting transactions—some transactions in this conflicting set are false positives. Kilo TM uses this approximate information to schedule hundreds of non-conflicting transactions for validation and commit in parallel. This approximate nature of recency bloom filter allows it to remain small, in the order of several kBs, and thus it can reside on-chip for fast access. Using the recency bloom filter to boost transaction commit parallelism is an integral part of Kilo TM.

Branch Divergence and Transactional Memory.   The transactional memory programming model introduces a new type of branch divergence. When a warp finishes a transaction, each of its active threads will try to commit. Some of the threads may abort and need to reexecute their transactions, while other threads may pass the validation and commit their transactions. Since this outcome may not be unanimous across the entire warp, a warp may diverge after validation. Fung et al. [2011] proposes a simple extension to the SIMT hardware to handle this specific kind of branch divergence introduced by transaction aborts. This extension is independent of other design aspects of Kilo TM, but it is a necessary piece for supporting TM on GPUs.

Figure 5.2 shows how the SIMT stack can be extended to handle control flow divergence due to transaction aborts. When a warp enters the transaction (at line B, tx_begin), it pushes two special entries onto the SIMT stack ❶. The first entry of type R stores information to restart the transaction. Its active mask is initially empty, and its PC field points to the instruction after tx_begin. The second entry of type T tracks the current transaction attempt. At tx_commit (line F), any thread that fails validation sets its mask bit in the R entry. The T entry is popped when the warp finishes the commit process (i.e., its active threads have either committed or aborted) ❷. A new T entry will then be pushed onto the stack using the active mask and PC from the R entry to restart the threads that have been aborted. Then, the active mask in the R entry is cleared ❸. If the active mask in the R entry is empty, both T and R entries are popped, revealing the original N entry ❺. Its PC is then modified to point to the instruction right after tx_commit, and the warp resumes normal execution. Branch divergence of a warp within a transaction is handled in the same way as non-transactional divergence ❹.

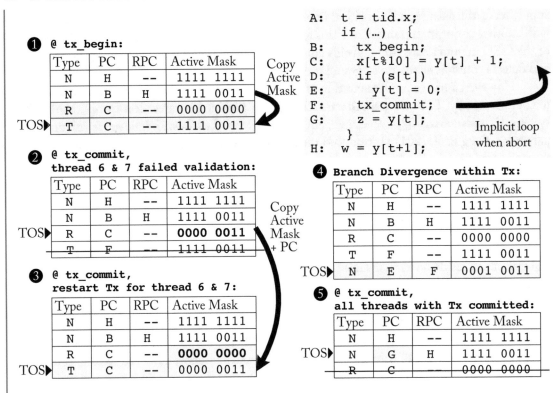

Figure 5.2: SIMT stack extension to handle divergence due to transaction aborts (validation fail). Thread 6 and 7 have failed validation and are restarted. Stack entry type: Normal (N), Transaction Retry (R), Transaction Top (T). For each scenario, added entries or modified fields are shaded.

5.3.2 WARP TM AND TEMPORAL CONFLICT DETECTION

In their follow-up paper, Fung and Aamodt [2013] propose two distinct enhancements that improves the performance and energy-efficiency of Kilo TM: warp-level transaction management (WarpTM) and temporal conflict detection (TCD).

Warp-level transaction management leverages the thread hierarchy in GPU programming models—the spatial locality among threads within a warp—to improve the efficiency of Kilo TM. In particular, WarpTM amortizes the control overhead of Kilo TM and boosts the utility of the GPU memory subsystem. These optimizations are only possible if conflicts within a warp can be resolved efficiently, and thus a low overhead intra-warp conflict resolution mechanism is crucial in maintaining the benefit from WarpTM. To this end, Fung and Aamodt [2013] propose a two-phase parallel intra-warp conflict resolution that resolves conflicts within a warp efficiently in parallel. With all intra-warp conflicts resolved, Kilo TM can merge the scalar memory accesses for the validation and commit of multiple transactions in the same warp into wider accesses. This optimization, called *validation and commit coalescing*, is key to enable Kilo TM to take full advantage of the wide GPU memory system, which is optimized for vector-wide accesses.

Temporal conflict detection is a low overhead mechanism that uses a set of globally synchronized on-chip timers to detect conflicts for read-only transactions. Once initialized, each of these on-chip timers runs locally in its microarchitecture module and does not communicate with other timers. This implicit synchronization without communication distinguishes TCD from existing timestamp-based conflict detections used in various software TM systems [Dalessandro et al., 2010, Spear et al., 2006, Xu et al., 2014]. TCD uses timestamps captured from these timers to infer the order of the memory reads of a transaction with respect to updates from other transactions. Kilo TM incorporates TCD to detect conflict-free read-only transactions that can commit directly without value-based conflict detection. In doing so, it significantly reduces the memory bandwidth overhead for these transactions, which can occur frequently for GPU-TM applications that use transactions for data structure traversals.

5.4 HETEROGENEOUS SYSTEMS

Concurrency Management in Heterogeneous Systems. Kayiran et al. [2014] propose a concurrency-limiting scheme to throttle GPU multithreading, reducing memory and network contention in multiprogrammed CPU/GPU systems. On heterogenous system, interference from the GPU can result in a significant performance drop for concurrently executing CPU applications. Their proposed thread-level-parallelism (TLP) limiting scheme observes congestion metrics in the shared CPU/GPU memory controller and interconnection network to estimate the number of GPU warps that should be actively scheduled on each GPU core. They propose two schemes, one that focuses on increasing CPU performance only and another that seeks to optimize overall system throughput (both CPU and GPU) by balancing GPU performance degradation due to constrained multithreading with CPU interference. The authors evaluate the performance impact of warp scheduling on a tiled heterogeneous architecture with a GPU core to CPU core ratio of 2:1, justified by an NVIDIA GPU SM being roughly half the area of a modern out-of-order Intel chip using the same process technology. The baseline configuration completely shares both the network bandwidth and memory controllers between the CPU and GPU, in the interest of maximizing resource utilization. Using this scheme, the authors make the observation that limiting GPU TLP can have both a positive and negative effect on GPU performance, but never hurts CPU performance.

To boost CPU performance, the authors introduce a CPU-centric concurrency management technique that monitors stalling in the global memory controllers. This technique separately counts the number of memory requests stalled due to the memory controller input queue being full, and the number of memory requests stalled because the reply network from the MC to the cores is full. These metrics are monitored locally at each memory controller and are aggregated in a centralized unit that sends the information to the GPU cores. Heuristically driven scheme set both high and low thresholds for these values. If the sum of both request stall counts is low (based on the threshold), then the number of warps actively scheduled on the GPU is increased. If the sum of both counts is high, then the number of active warps is decreased in hopes that CPU performance will increase as a result of less GPU memory traffic.

A more balanced technique that attempts to maximize overall system throughput augments the CPU-centric approach to account for the impact on GPU performance from warp throttling. This balanced technique monitors the number of cycles the GPU cannot issue for during a concurrency rebalancing interval (1,024 cycles in their work). The moving average of GPU stalling for the current multithreading limit level is stored on each GPU core and used to determine if the level of multithreading should be increased or decreased. The balanced technique modulates the GPU's TLP in two phases. In the first phase its operation is identical to the CPU-centric solution, where GPU stalling is not accounted for and GPU TLP is limited based only on memory contention. In the second phase (which begins once GPU TLP throttling starts to cause a GPU performance degradation because the GPU's tolerance of latency has been reduced), the system stops throttling GPU concurrency if it predicts that doing so will harm GPU performance. This prediction is made by looking up the moving average number of GPU stalls at the target multithreading level, which has been recorded from earlier executions at this level. If the difference between observed GPU stalls at the target multithreading level and the current multithreading level exceeds a threshold k value, then the TLP level is not decreased. This k value can be set by the user and is a proxy for specifying the priority of GPU performance.

Heterogeneous System Coherence.   Power et al. [2013a] propose a hardware mechanism to efficiently support cache coherence between the CPU and the GPU on an integrated system. They identify that directory bandwidth is significant bottleneck with the increased memory traffic generated by the GPU. They employ coarse-grained region coherence [Cantin et al., 2005] to cut down on excessive directory traffic caused in a traditional cache block-based coherence directory. Once permissions for the coarse-grained region have been acquired, most requests will not have to access the directory and the coherence traffic can be offloaded to an incoherent direct-access bus instead of a lower-bandwidth coherent interconnection network.

Heterogeneous TLP-Aware Cache Management for CPU-GPU Architectures.   Lee and Kim [2012] evaluate the effect managing a shared last-level cache (LLC) between CPU cores and GPU cores in a heterogeneous environment. They demonstrate that while cache hit rates are a key performance metric for CPU workloads, many GPU workloads are insensitive to cache hit rate, since memory latency can be hidden by thread-level parallelism. To determine if GPU apps are cache-sensitive, they develop a per-core performance sampling technique where some cores bypass the shared LLC and some insert at the most-recently used position. Based on the relative performance of these cores, they can set the bypassing policy for the rest of the GPU cores, inserting in the LLC if performance is improved and bypassing if they are insensitive.

Second, they observe that previous CPU-centric cache management favors cores with more frequent accesses. GPU cores are shown to generate five to ten times more traffic at the LLC. This increases bias cache capacity toward the GPU, decreasing the performance of CPU apps. They propose extending previously proposed work on utility-based cache partitioning [Qureshi and Patt, 2006] to account for the relative ratio of LLC accesses. When GPU cores are cache-sensitive, the CPU core's accesses cache way-allocation is increased beyond what utility-based cache partitioning provides to account for the difference in access magnitude and latency sensitivity between CPUs and GPUs.

Bibliography

Die photo analysis. `http://vlsiarch.eecs.harvard.edu/accelerators/die-photo-analysis` 1

`https://en.wikipedia.org/wiki/GDDR5_SDRAM` 76

`Top500.org` 2

Tor M. Aamodt, Wilson W. L. Fung, Inderpreet Singh, Ahmed El-Shafiey, Jimmy Kwa, Tayler Hetherington, Ayub Gubran, Andrew Boktor, Tim Rogers, Ali Bakhoda, and Hadi Jooybar. *GPGPU-Sim 3.x Manual.* 16, 38

M. Abdel-Majeed and M. Annavaram. Warped register file: A power efficient register file for GPGPUs. In *Proc. of the IEEE International Symposium on High-Performance Computer Architecture (HPCA)*, February 2013. DOI: 10.1109/hpca.2013.6522337. 64

M. Abdel-Majeed, A. Shafaei, H. Jeon, M. Pedram, and M. Annavaram. Pilot register file: Energy efficient partitioned register file for GPUs. In *Proc. of the IEEE International Symposium on High-Performance Computer Architecture (HPCA)*, February 2017. DOI: 10.1109/hpca.2017.47. 65

Dominic Acocella and Mark R. Goudy. United States Patent #7,750,915: Concurrent Access of Data Elements Stored Across Multiple Banks in a Shared Memory Resource (Assignee: NVIDIA Corp.), July 2010. 68

Neha Agarwal, David Nellans, Mark Stephenson, Mike O'Connor, and Stephen W. Keckler. Page placement strategies for GPUs within heterogeneous memory systems. In *Proc. of the ACM Architectural Support for Programming Languages and Operating Systems (ASPLOS)*, 2015. DOI: 10.1145/2694344.2694381. 82

Jade Alglave, Mark Batty, Alastair F. Donaldson, Ganesh Gopalakrishnan, Jeroen Ketema, Daniel Poetzl, Tyler Sorensen, and John Wickerson. GPU concurrency: Weak behaviours and programming assumptions. In *Proc. of the ACM Architectural Support for Programming Languages and Operating Systems (ASPLOS)*, pages 577–591, 2015. DOI: 10.1145/2694344.2694391. 72

John R. Allen, Ken Kennedy, Carrie Porterfield, and Joe Warren. Conversion of control dependence to data dependence. In *Proc. of the ACM Symposium on Principles and Practices of Parallel Programming (PPoPP)*, pages 177–189, 1983. DOI: 10.1145/567067.567085. 16

Robert Alverson, David Callahan, Daniel Cummings, Brian Koblenz, Allan Porterfield, and Burton Smith. The tera computer system. In *Proc. of the ACM International Conference on Supercomputing (ICS)*, pages 1–6, 1990. DOI: 10.1145/77726.255132. 16

R700-Family Instruction Set Architecture. AMD, March 2009. 49

AMD Southern Islands Series Instruction Set Architecture. AMD, 1.1 ed., December 2012. 13, 17, 19, 20, 26, 54, 58, 74

A. Arunkumar, S. Y. Lee, and C. J. Wu. ID-cache: Instruction and memory divergence based cache management for GPUs. In *Proc. of the IEEE International Symposium on Workload Characterization (IISWC)*, 2016. DOI: 10.1109/iiswc.2016.7581276. 79

Akhil Arunkumar, Evgeny Bolotin, Benjamin Cho, Ugljesa Milic, Eiman Ebrahimi, Oreste Villa, Aamer Jaleel, Carole-Jean Wu, and David Nellans. MCM-GPU: Multi-chip-module GPUs for continued performance scalability. In *Proc. of the ACM/IEEE International Symposium on Computer Architecture (ISCA)*, pages 320–332, 2017. DOI: 10.1145/3079856.3080231. 84

Krste Asanovic, Stephen W. Keckler, Yunsup Lee, Ronny Krashinsky, and Vinod Grover. Convergence and scalarization for data-parallel architectures. In *Proc. of the ACM/IEEE International Symposium on Code Generation and Optimization (CGO)*, 2013. DOI: 10.1109/cgo.2013.6494995. 54, 56, 58

Rachata Ausavarungnirun, Saugata Ghose, Onur Kayran, Gabriel H. Loh, Chita R. Das, Mahmut T. Kandemir, and Onur Mutlu. Exploiting inter-warp heterogeneity to improve GPGPU performance. In *Proc. of the ACM/IEEE International Conference on Parallel Architecture and Compilation Techniques (PACT)*, 2015. DOI: 10.1109/pact.2015.38. 80

Ali Bakhoda, George L. Yuan, Wilson W. L. Fung, Henry Wong, and Tor M. Aamodt. Analyzing CUDA workloads using a detailed GPU simulator. In *Proc. of the IEEE Symposium of Performance and Analysis of Systems and Software, (ISPASS'09)*, pages 163–174, 2009. DOI: 10.1109/ispass.2009.4919648. 6, 14, 78

Ali Bakhoda, John Kim, and Tor M. Aamodt. Throughput-effective on-chip networks for many-core accelerators. In *Proc. of the ACM/IEEE International Symposium on Microarchitecture (MICRO)*, pages 421–432, 2010. DOI: 10.1109/micro.2010.50. 77

Ali Bakhoda, John Kim, and Tor M. Aamodt. Designing on-chip networks for throughput accelerators. *ACM Transactions on Architecture and Code Optimization (TACO)*, 10(3):21, 2013. DOI: 10.1145/2509420.2512429. 77

Markus Billeter, Ola Olsson, and Ulf Assarsson. Efficient stream compaction on wide SIMD many-core architectures. In *Proc. of the ACM Conference on High Performance Graphics*, pages 159–166, 2009. DOI: 10.1145/1572769.1572795. 45

Nicolas Brunie, Sylvain Collange, and Gregory Diamos. Simultaneous branch and warp interweaving for sustained GPU performance. In *Proc. of the ACM/IEEE International Symposium on Computer Architecture (ISCA)*, pages 49–60, 2012. DOI: 10.1109/isca.2012.6237005. 44, 53

Ian Buck, Tim Foley, Daniel Horn, Jeremy Sugerman, Kayvon Fatahalian, Mike Houston, and Pat Hanrahan. Brook for GPUs: Stream computing on graphics hardware. In *Proc. of the ACM International Conference on Computer Graphics and Interactive Techniques (SIGGRAPH)*, pages 777–786, 2004. DOI: 10.1145/1186562.1015800. 6

Brian Cabral. what is "SASS" short for? https://stackoverflow.com/questions/9798258/what-is-sass-short-for, November 2016. 14

J. F. Cantin, M. H. Lipasti, and J. E. Smith. Improving multiprocessor performance with coarse-grain coherence tracking. In *Proc. of the ACM/IEEE International Symposium on Computer Architecture (ISCA)*, June 2005. DOI: 10.1109/isca.2005.31. 100

Edwin Catmull. A subdivision algorithm for computer display of curved surfaces. *Technical Report*, DTIC Document, 1974. 72

Hassan Chafi, Jared Casper, Brian D. Carlstrom, Austen Mcdonald, Chi Cao, Minh Woongki Baek, Christos Kozyrakis, and Kunle Olukotun. A scalable, non-blocking approach to transactional memory. In *Proc. of the IEEE International Symposium on High-Performance Computer Architecture (HPCA)*, pages 97–108, 2007. DOI: 10.1109/hpca.2007.346189. 97

Guoyang Chen, Bo Wu, Dong Li, and Xipeng Shen. PORPLE: An extensible optimizer for portable data placement on GPU. In *Proc. of the ACM/IEEE International Symposium on Microarchitecture (MICRO)*, 2014a. DOI: 10.1109/micro.2014.20. 83

Xi E. Chen and Tor M. Aamodt. A first-order fine-grained multithreaded throughput model. In *Proc. of the IEEE International Symposium on High-Performance Computer Architecture (HPCA)*, pages 329–340, 2009. DOI: 10.1109/hpca.2009.4798270. 78

Xuhao Chen, Li-Wen Chang, Christopher I. Rodrigues, Jie Lv, Zhiying Wang, and Wen-Mei Hwu. Adaptive cache management for energy-efficient GPU computing. In *Proc. of the ACM/IEEE International Symposium on Microarchitecture (MICRO)*, 2014b. DOI: 10.1109/micro.2014.11. 81, 82

Sylvain Collange, David Defour, and Yao Zhang. Dynamic detection of uniform and affine vectors in GPGPU computations. In *Proc. of the European Conference on Parallel Processing (Euro-Par)*, 2010. DOI: 10.1007/978-3-642-14122-5_8. 59

Brett W. Coon and John Erik Lindholm. United States Patent #7,353,369: System and Method for Managing Divergent Threads in a SIMD Architecture (Assignee NVIDIA Corp.), April 2008. 26, 49, 50

Brett W. Coon, Peter C. Mills, Stuart F. Oberman, and Ming Y. Siu. United States Patent #7,434,032: Tracking Register Usage During Multithreaded Processing Using a Scoreboard having Separate Memory Regions and Storing Sequential Register Size Indicators (Assignee NVIDIA Corp.), October 2008. 34

Brett W. Coon, John Erik Lindholm, Gary Tarolli, Svetoslav D. Tzvetkov, John R. Nickolls, and Ming Y. Siu. United States Patent #7,634,621: Register File Allocation (Assignee NVIDIA Corp.), December 2009. 35

Ron Cytron, Jeanne Ferrante, Barry K. Rosen, Mark N. Wegman, and F. Kenneth Zadeck. Efficiently computing static single assignment form and the control dependence graph. *ACM Transactions on Programming Languages and Systems (TOPLAS)*, 13(4):451–490, 1991. DOI: 10.1145/115372.115320. 16

Luke Dalessandro, Michael F. Spear, and Michael L. Scott. NOrec: Streamlining STM by abolishing ownership records. In *Proc. of the ACM Symposium on Principles and Practices of Parallel Programming (PPoPP)*, pages 67–78, 2010. DOI: 10.1145/1693453.1693464. 96, 99

R. H. Dennard, F. H. Gaensslen, and K. Mai. Design of ion-implanted MOSFET's with very small physical dimensions. *IEEE Journal of Solid-State Circuits*, October 1974. DOI: 10.1109/jssc.1974.1050511. 1

Gregory Diamos, Benjamin Ashbaugh, Subramaniam Maiyuran, Andrew Kerr, Haicheng Wu, and Sudhakar Yalamanchili. SIMD re-convergence at thread frontiers. In *Proc. of the ACM/IEEE International Symposium on Microarchitecture (MICRO)*, pages 477–488, 2011. DOI: 10.1145/2155620.2155676. 26, 51, 54

Gregory Frederick Diamos, Richard Craig Johnson, Vinod Grover, Olivier Giroux, Jack H. Choquette, Michael Alan Fetterman, Ajay S. Tirumala, Peter Nelson, and Ronny Meir Krashinsky. Execution of divergent threads using a convergence barrier, July 13, 2015. 27, 28, 29, 30

Roger Eckert. United States Patent #7,376,803: Page stream sorter for DRAM systems (Assignee: NVIDIA Corp.), May 2008. 73

Roger Eckert. United States Patent #9,195,618: Method and System for Scheduling Memory Requests (Assignee: NVIDIA Corp.), November 2015. 73

John H. Edmondson and James M. Van Dyke. United States Patent #7872657: Memory Addressing Scheme using Partition Strides, January 2011. 75

John H. Edmondson et al. United States Patent #8,464,001: Cache and Associated Method with Frame Buffer Managed Dirty Data Pull and High-Priority Clean Mechanism, June 2013. 75, 76

Ahmed ElTantaway, Jessica Wenjie Ma, Mike O'Connor, and Tor M. Aamodt. A scalable multi-path microarchitecture for efficient GPU control flow. In *Proc. of the IEEE International Symposium on High-Performance Computer Architecture (HPCA)*, 2014. DOI: 10.1109/hpca.2014.6835936. 26, 28, 30, 31, 49, 52

Ahmed ElTantawy and Tor M. Aamodt. MIMD synchronization on SIMT architectures. In *Proc. of the ACM/IEEE International Symposium on Microarchitecture (MICRO)*, pages 1–14, 2016. DOI: 10.1109/micro.2016.7783714. 26, 27, 30, 32, 93

Ahmed ElTantawy and Tor M. Aamodt. Warp scheduling for fine-grained synchronization. In *Proc. of the IEEE International Symposium on High-Performance Computer Architecture (HPCA)*, pages 375–388, 2018. DOI: 10.1109/hpca.2018.00040. 93

Alexander L. Minken et al., United States Patent #7,649,538: Reconfigurable High Performance Texture Pipeline with Advanced Filtering (Assignee: NVIDIA Corp.), January 2010. 72

Wilson W. L. Fung. *GPU Computing Architecture for Irregular Parallelism*. Ph.D. thesis, University of British Columbia, January 2015. DOI: 10.14288/1.0167110. 41

Wilson W. L. Fung and Tor M. Aamodt. Thread block compaction for efficient SIMT control flow. In *Proc. of the IEEE International Symposium on High-Performance Computer Architecture (HPCA)*, pages 25–36, 2011. DOI: 10.1109/hpca.2011.5749714. 26, 28, 42, 43, 50, 91

Wilson W. L. Fung and Tor M. Aamodt. Energy efficient GPU transactional memory via spacetime optimizations. In *Proc. of the ACM/IEEE International Symposium on Microarchitecture (MICRO)*, pages 408–420, 2013. DOI: 10.1145/2540708.2540743. 98

Wilson W. L. Fung, Ivan Sham, George Yuan, and Tor M. Aamodt. Dynamic warp formation and scheduling for efficient GPU control flow. In *Proc. of the ACM/IEEE International Symposium on Microarchitecture (MICRO)*, pages 407–420, 2007. DOI: 10.1109/micro.2007.4408272. 14, 23, 25, 42, 44, 49, 91

Wilson W. L. Fung, Inderpreet Singh, Andrew Brownsword, and Tor M. Aamodt. Hardware transactional memory for GPU architectures. In *Proc. of the ACM/IEEE International Symposium on Microarchitecture (MICRO)*, pages 296–307, 2011. DOI: 10.1145/2155620.2155655. 96, 97

Wilson Fung et al. Dynamic warp formation: Efficient MIMD control flow on SIMD graphics hardware. *ACM Transactions on Architecture and Code Optimization (TACO)*, 6(2):7:1–7:37, 2009. DOI: 10.1145/1543753.1543756. 42, 49

M. Gebhart, S. W. Keckler, and W. J. Dally. A compile-time managed multi-level register file hierarchy. In *Proc. of the ACM/IEEE International Symposium on Microarchitecture (MICRO)*, December 2011a. DOI: 10.1145/2155620.2155675. 63

Mark Gebhart, Daniel R. Johnson, David Tarjan, Stephen W. Keckler, William J. Dally, Erik Lindholm, and Kevin Skadron. Energy-efficient mechanisms for managing thread context in throughput processors. In *Proc. of the ACM/IEEE International Symposium on Computer Architecture (ISCA)*, 2011b. DOI: 10.1145/2000064.2000093. 63

Mark Gebhart, Daniel R. Johnson, David Tarjan, Stephen W. Keckler, William J. Dally, Erik Lindholm, and Kevin Skadron. Energy-efficient mechanisms for managing thread context in throughput processors. In *Proc. of the ACM/IEEE International Symposium on Computer Architecture (ISCA)*, pages 235–246, 2011c. DOI: 10.1145/2000064.2000093. 88

Isaac Gelado, John E. Stone, Javier Cabezas, Sanjay Patel, Nacho Navarro, and Wen-mei W. Hwu. An asymmetric distributed shared memory model for heterogeneous parallel systems. In *Proc. of the ACM Architectural Support for Programming Languages and Operating Systems (ASPLOS)*, pages 347–358, 2010. DOI: 10.1145/1736020.1736059. 3

Syed Zohaib Gilani, Nam Sung Kim, and Michael J. Schulte. Power-efficient computing for compute-intensive GPGPU applications. In *Proc. of the IEEE International Symposium on High-Performance Computer Architecture (HPCA)*, 2013. DOI: 10.1109/hpca.2013.6522330. 59, 61

David B. Glasco et al. United States Patent #8,135,926: Cache-Based Control of Atomic Operations in Conjunction with an External Alu Block, March 2012. 76

David B. Glasco et al. United States Patent #8,539,130: Virtual Channels for Effective Packet Transfer, September 2013. 75

Scott Gray. Assembler for NVIDIA Maxwell architecture. https://github.com/NervanaSystems/maxas 16, 40

Zvika Guz, Evgeny Bolotin, Idit Keidar, Avinoam Kolodny, Avi Mendelson, and Uri C. Weiser. Many-core vs. many-thread machines: Stay away from the valley. *IEEE Computer Architecture Letters*, 8(1):25–28, 2009. DOI: 10.1109/l-ca.2009.4. 5

Ziyad S. Hakura and Anoop Gupta. The design and analysis of a cache architecture for texture mapping. In *Proc. of the ACM/IEEE International Symposium on Computer Architecture (ISCA)*, pages 108–120, 1997. DOI: 10.1145/264107.264152. 73

Rehan Hameed, Wajahat Qadeer, Megan Wachs, Omid Azizi, Alex Solomatnikov, Benjamin C. Lee, Stephen Richardson, Christos Kozyrakis, and Mark Horowitz. Understanding sources of inefficiency in general-purpose chips. In *Proc. of the ACM/IEEE International Symposium on Computer Architecture (ISCA)*, pages 37–47, 2010. DOI: 10.1145/1815961.1815968. 1

Song Han, Xingyu Liu, Huizi Mao, Jing Pu, Ardavan Pedram, Mark A. Horowitz, and William J. Dally. EIE: efficient inference engine on compressed deep neural network. In

Proc. of the ACM/IEEE International Symposium on Computer Architecture (ISCA), pages 243–254, 2016. DOI: 10.1109/isca.2016.30. 6

Mark Harris. *An Easy Introduction to CUDA C and C++.* https://devblogs.nvidia.com/parallelforall/easy-introduction-cuda-c-and-c/, 2012. 11

Tim Harris, James Larus, and Ravi Rajwar. *Transactional Memory*, 2nd ed. Morgan & Claypool, 2010. DOI: 10.1201/b11417-16. 96

Steven J. Heinrich et al. United States Patent #9,595,075: Load/Store Operations in Texture Hardware (Assignee: NVIDIA Corp.), March 2017. 68, 73

John Hennessy and David Patterson. *Computer Architecture—A Quantitative Approach*, 5th ed. Morgan Kaufmann, 2011. 1, 10, 71, 72, 78

Maurice Herlihy and J. Eliot B. Moss. Transactional memory: Architectural support for lock-free data structures. In *Proc. of the ACM/IEEE International Symposium on Computer Architecture (ISCA)*, pages 289–300, 1993. DOI: 10.1109/isca.1993.698569. 96

Jared Hoberock, Victor Lu, Yuntao Jia, and John C. Hart. Stream compaction for deferred shading. In *Proc. of the ACM Conference on High Performance Graphics*, pages 173–180, 2009. DOI: 10.1145/1572769.1572797. 45

H. Peter Hofstee. Power efficient processor architecture and the cell processor. In *Proc. of the IEEE International Symposium on High-Performance Computer Architecture (HPCA)*, pages 258–262, 2005. DOI: 10.1109/hpca.2005.26. 68

Mark Horowitz, Elad Alon, Dinesh Patil, Samuel Naffziger, Rajesh Kumar, and Kerry Bernstein. Scaling, power, and the future of CMOS. In *IEEE International Electron Devices Meeting*, 2005. DOI: 10.1109/iedm.2005.1609253. 5

Homan Igehy, Matthew Eldridge, and Kekoa Proudfoot. Prefetching in a texture cache architecture. In *Proc. of the ACM SIGGRAPH/EUROGRAPHICS Workshop on Graphics hardware*, pages 133–ff, 1998. DOI: 10.1145/285305.285321. 72, 73, 74

Hyeran Jeon, Gokul Subramanian Ravi, Nam Sung Kim, and Murali Annavaram. GPU register file virtualization. In *Proc. of the ACM/IEEE International Symposium on Microarchitecture (MICRO)*, pages 420–432, 2015. DOI: 10.1145/2830772.2830784. 64

W. Jia, K. A. Shaw, and M. Martonosi. MRPB: Memory request prioritization for massively parallel processors. In *Proc. of the IEEE International Symposium on High-Performance Computer Architecture (HPCA)*, 2014. DOI: 10.1109/hpca.2014.6835938. 78, 79

Wenhao Jia, Kelly A Shaw, and Margaret Martonosi. Characterizing and improving the use of demand-fetched caches in GPUs. In *Proc. of the ACM International Conference on Supercomputing (ICS)*, pages 15–24, 2012. DOI: 10.1145/2304576.2304582. 78

Adwait Jog, Onur Kayiran, Nachiappan Chidambaram Nachiappan, Asit K. Mishra, Mahmut T. Kandemir, Onur Mutlu, Ravishankar Iyer, and Chita R. Das. OWL: Cooperative thread array aware scheduling techniques for improving GPGPU performance. In *Proc. of the ACM Architectural Support for Programming Languages and Operating Systems (ASPLOS)*, 2013a. DOI: 10.1145/2451116.2451158. 90

Adwait Jog, Onur Kayiran, Asit K. Mishra, Mahmut T. Kandemir, Onur Mutlu, Ravishankar Iyer, and Chita R. Das. Orchestrated scheduling and prefetching for GPGPUs. In *Proc. of the ACM/IEEE International Symposium on Computer Architecture (ISCA)*, 2013b. DOI: 10.1145/2508148.2485951. 90

Norman P. Jouppi, Cliff Young, Nishant Patil, David Patterson, Gaurav Agrawal, Raminder Bajwa, Sarah Bates, Suresh Bhatia, Nan Boden, Al Borchers, et al. In-datacenter performance analysis of a tensor processing unit. In *Proc. of the ACM/IEEE International Symposium on Computer Architecture (ISCA)*, 2017. DOI: 10.1145/3079856.3080246. 2

David R. Kaeli, Perhaad Mistry, Dana Schaa, and Dong Ping Zhang. *Heterogeneous Computing with OpenCL 2.0*. Morgan Kaufmann, 2015. 10

Ujval J. Kapasi et al. Efficient conditional operations for data-parallel architectures. In *Proc. of the ACM/IEEE International Symposium on Microarchitecture (MICRO)*, pages 159–170, 2000. DOI: 10.1109/micro.2000.898067. 45

O. Kayiran, N. C. Nachiappan, A. Jog, R. Ausavarungnirun, M. T. Kandemir, G. H. Loh, O. Mutlu, and C. R. Das. Managing GPU concurrency in heterogeneous architectures. In *Proc. of the ACM/IEEE International Symposium on Microarchitecture (MICRO)*, 2014. DOI: 10.1109/micro.2014.62. 99

Onur Kayiran, Adwait Jog, Mahmut T. Kandemir, and Chita R. Das. Neither more nor less: Optimizing thread-level parallelism for GPGPUs. In *Proc. of the ACM/IEEE International Conference on Parallel Architecture and Compilation Techniques (PACT)*, 2013. 86

S. W. Keckler, W. J. Dally, B. Khailany, M. Garland, and D. Glasco. GPUs and the future of parallel computing. *Micro, IEEE*, 31(5):7–17, September 2011. DOI: 10.1109/mm.2011.89. 53, 58

Shane Keil and John H. Edmondson. United States Patent #8,195,858: Managing Conflicts on Shared L2 Bus, June 2012. 77

Shane Keil et al. United States Patent #8,307,165: Sorting Requests to the Dram for High Page Locality, November 2012. 77

Farzad Khorasani, Rajiv Gupta, and Laxmi N. Bhuyan. Efficient warp execution in presence of divergence with collaborative context collection. In *Proc. of the ACM/IEEE International Symposium on Microarchitecture (MICRO)*, 2015. DOI: 10.1145/2830772.2830796. 45

J. Y. Kim and C. Batten. Accelerating irregular algorithms on GPGPUs using fine-grain hardware worklists. In *Proc. of the ACM/IEEE International Symposium on Microarchitecture (MICRO)*, 2014. DOI: 10.1109/micro.2014.24. 93

Ji Kim, Christopher Torng, Shreesha Srinath, Derek Lockhart, and Christopher Batten. Microarchitectural mechanisms to exploit value structure in SIMT architectures. In *Proc. of the ACM/IEEE International Symposium on Computer Architecture (ISCA)*, 2013. DOI: 10.1145/2508148.2485934. 57, 58, 59, 60, 61, 62

Sangman Kim, Seonggu Huh, Xinya Zhang, Yige Hu, Amir Wated, Emmett Witchel, and Mark Silberstein. GPUnet: Networking abstractions for GPU programs. In *Proc. of the USENIX Symposium on Operating Systems Design and Implementation*, pages 6–8, 2014. DOI: 10.1145/2963098. 2

David B. Kirk and W. Hwu Wen-Mei. *Programming Massively Parallel Processors: A Hands-on Approach*. Morgan Kaufmann, 2016. DOI: 10.1016/c2011-0-04129-7. 9

John Kloosterman, Jonathan Beaumont, D. Anoushe Jamshidi, Jonathan Bailey, Trevor Mudge, and Scott Mahlke. Regless: Just-in-time operand staging for GPUs. In *Proc. of the ACM/IEEE International Symposium on Microarchitecture (MICRO)*, pages 151–164, 2017. DOI: 10.1145/3123939.3123974. 65

R. Krashinsky, C. Batten, M. Hampton, S. Gerding, B. Pharris, J. Casper, and K. Asanovic. The vector-thread architecture. In *Proc. of the ACM/IEEE International Symposium on Computer Architecture (ISCA)*, pages 52–63, June 2004. DOI: 10.1109/isca.2004.1310763. 52

Ronny M. Krashinsky. United States Patent Application #20130042090 A1: Temporal SIMT Execution Optimization, August 2011. 53, 58

David Kroft. Lockup-free instruction fetch/prefetch cache organization. In *Proc. of the ACM/IEEE International Symposium on Computer Architecture (ISCA)*, pages 81–87, 1981. DOI: 10.1145/285930.285979. 33, 71

Jens Krüger and Rüdiger Westermann. Linear algebra operators for GPU implementation of numerical algorithms. In *ACM Transactions on Graphics (TOG)*, V. 22, pages 908–916, 2003. DOI: 10.1145/882262.882363. 6

Junjie Lai and André Seznec. Performance upper bound analysis and optimization of SGEMM on Fermi and Kepler GPUs. In *Proc. of the ACM/IEEE International Symposium on Code Generation and Optimization (CGO)*, pages 1–10, 2013. DOI: 10.1109/cgo.2013.6494986. 16

Nagesh B. Lakshminarayana and Hyesoon Kim. Effect of instruction fetch and memory scheduling on GPU performance. In *Workshop on Language, Compiler, and Architecture Support for GPGPU*, 2010. 88

Ahmad Lashgar, Ebad Salehi, and Amirali Baniasadi. A case study in reverse engineering GPG-PUs: Outstanding memory handling resources. *ACM SIGARCH Computer Architecture News*, 43(4):15–21, 2016. DOI: 10.1145/2927964.2927968. 34

C. L. Lawson, R. J. Hanson, D. R. Kincaid, and F. T. Krogh. Basic linear algebra subprograms for fortran usage. *ACM Transactions on Mathematical Software*, 5(3):308–323, September 1979. DOI: 10.1145/355841.355848. 10

HyoukJoong Lee, Kevin J. Brown, Arvind K. Sujeeth, Tiark Rompf, and Kunle Olukotun. Locality-aware mapping of nested parallel patterns on GPUs. In *Proc. of the ACM/IEEE International Symposium on Microarchitecture (MICRO)*, 2014a. DOI: 10.1109/micro.2014.23. 94

J. Lee and H. Kim. TAP: A TLP-aware cache management policy for a CPU-GPU heterogeneous architecture. In *Proc. of the IEEE International Symposium on High-Performance Computer Architecture (HPCA)*, 2012. DOI: 10.1109/hpca.2012.6168947. 100

S. Y. Lee and C. J. Wu. Ctrl-C: Instruction-aware control loop based adaptive cache bypassing for GPUs. In *Proc. of the IEEE International Conference on Computer Design (ICCD)*, pages 133–140, 2016. DOI: 10.1109/iccd.2016.7753271. 80

Sangpil Lee, Keunsoo Kim, Gunjae Koo, Hyeran Jeon, Won Woo Ro, and Murali Annavaram. Warped-compression: Enabling power efficient GPUs through register compression. In *Proc. of the ACM/IEEE International Symposium on Computer Architecture (ISCA)*, 2015. DOI: 10.1145/2749469.2750417. 59, 60, 61

Shin-Ying Lee and Carole-Jean Wu. CAWS: Criticality-aware warp scheduling for GPGPU workloads. In *Proc. of the ACM/IEEE International Conference on Parallel Architecture and Compilation Techniques (PACT)*, 2014. DOI: 10.1145/2628071.2628107. 93

Victor W. Lee, Changkyu Kim, Jatin Chhugani, Michael Deisher, Daehyun Kim, Anthony D. Nguyen, Nadathur Satish, Mikhail Smelyanskiy, Srinivas Chennupaty, Per Hammarlund, et al. Debunking the 100X GPU vs. CPU myth: An evaluation of throughput computing on CPU and GPU. In *Proc. of the ACM/IEEE International Symposium on Computer Architecture (ISCA)*, pages 451–460, 2010. DOI: 10.1145/1815961.1816021. 2

Yunsup Lee, Rimas Avizienis, Alex Bishara, Richard Xia, Derek Lockhart, Christopher Batten, and Krste Asanović. Exploring the tradeoffs between programmability and efficiency in data-parallel accelerators. In *Proc. of the ACM/IEEE International Symposium on Computer Architecture (ISCA)*, pages 129–140, 2011. DOI: 10.1145/2000064.2000080. 53

Yunsup Lee, Vinod Grover, Ronny Krashinsky, Mark Stephenson, Stephen W. Keckler, and Krste Asanović. Exploring the design space of SPMD divergence management on data-parallel architectures. In *Proc. of the ACM/IEEE International Symposium on Microarchitecture (MICRO)*, 2014b. DOI: 10.1109/micro.2014.48. 55, 56

Jingwen Leng, Tayler Hetherington, Ahmed ElTantawy, Syed Gilani, Nam Sung Kim, Tor M. Aamodt, and Vijay Janapa Reddi. GPUWattch: Enabling energy optimizations in GPG-PUs. In *Proc. of the ACM/IEEE International Symposium on Computer Architecture (ISCA)*, pages 487–498, 2013. DOI: 10.1145/2508148.2485964. 6

Adam Levinthal and Thomas Porter. Chap—A SIMD graphics processor. In *Proc. of the ACM International Conference on Computer Graphics and Interactive Techniques (SIGGRAPH)*, pages 77–82, 1984. DOI: 10.1145/800031.808581. 50

Dong Li, Minsoo Rhu, Daniel R. Johnson, Mike O'Connor, Mattan Erez, Doug Burger, Donald S. Fussell, and Stephen W. Redder. Priority-based cache allocation in throughput processors. In *Proc. of the IEEE International Symposium on High-Performance Computer Architecture (HPCA)*, 2015. DOI: 10.1109/hpca.2015.7056024. 82

E. Lindholm, J. Nickolls, S. Oberman, and J. Montrym. NVIDIA Tesla: A unified graphics and computing architecture. *Micro, IEEE*, 28(2):39–55, March–April 2008a. DOI: 10.1109/mm.2008.31. 9

Erik Lindholm, Mark J. Kilgard, and Henry Moreton. A user-programmable vertex engine. In *Proc. of the ACM International Conference on Computer Graphics and Interactive Techniques (SIGGRAPH)*, pages 149–158, 2001. DOI: 10.1145/383259.383274. 6, 21

John Erik Lindholm, Ming Y. Siu, Simon S. Moy, Samuel Liu, and John R. Nickolls. United States Patent #7,339,592: Simulating Multiported Memories Using Lower Port Count Memories (Assignee NVIDIA Corp.), March 2008b. 35, 38

Erik Lindholm et al. United States Patent #9,189,242: Credit-Based Streaming Multiprocessor Warp Scheduling (Assignee NVIDIA Corp.), November 2015. 33, 41

John S. Liptay. Structural aspects of the system/360 model 85, II: The cache. *IBM Systems Journal*, 7(1):15–21, 1968. DOI: 10.1147/sj.71.0015. 70

Z. Liu, S. Gilani, M. Annavaram, and N. S. Kim. G-Scalar: Cost-effective generalized scalar execution architecture for power-efficient GPUs. In *Proc. of the IEEE International Symposium on High-Performance Computer Architecture (HPCA)*, 2017. DOI: 10.1109/hpca.2017.51. 60, 61, 62

Samuel Lui, John Erik Lindholm, Ming Y. Siu, Brett W. Coon, and Stuart F. Oberman. United States Patent Application 11/555,649: Operand Collector Architecture (Assignee NVIDIA Corp.), May 2008. 35

Michael D. McCool, Arch D. Robison, and James Reinders. *Structured Parallel Programming: Patterns for Efficient Computation.* Elsevier, 2012. 10

Jiayuan Meng, David Tarjan, and Kevin Skadron. Dynamic warp subdivision for integrated branch and memory divergence tolerance. In *Proc. of the ACM/IEEE International Symposium on Computer Architecture (ISCA)*, pages 235–246, 2010. DOI: 10.1145/1815961.1815992. 30, 48, 90

Alexander L. Minken and Oren Rubinstein. United States Patent #6,629,188: Circuit and Method for Prefetching Data for a Texture Cache (Assignee: NVIDIA Corp.), September 2003. 72

Alexander L. Minkin et al. United States Patent #8,266,383: Cache Miss Processing Usinga Defer/Replay Mechanism (Assignee: NVIDIA Corp.), September 2012. 68, 69, 71

Alexander L. Minkin et al. United States Patent #8,595,425: Configurable Cache for Multiple Clients (Assignee: NVIDIA Corp.), November 2013. 68

Michael Mishkin, Nam Sung Kim, and Mikko Lipasti. Write-after-read hazard prevention in GPGPUSIM. In *Workshop on Deplicating, Deconstructing, and Debunking (WDDD)*, June 2016. 39, 40

John Montrym and Henry Moreton. The geforce 6800. *IEEE Micro*, 25(2):41–51, 2005. DOI: 10.1109/mm.2005.37. 1

Veynu Narasiman, Michael Shebanow, Chang Joo Lee, Rustam Miftakhutdinov, Onur Mutlu, and Yale N. Patt. Improving GPU performance via large warps and two-level warp scheduling. In *Proc. of the ACM/IEEE International Symposium on Microarchitecture (MICRO)*, pages 308–317, 2011. DOI: 10.1145/2155620.2155656. 33, 38, 43, 88, 91

John R. Nickolls and Jochen Reusch. Autonomous SIMD flexibility in the MP-1 and MP-2. In *Proc. of the ACM Symposium on Parallel Algorithms and Architectures (SPAA)*, pages 98–99, 1993. DOI: 10.1145/165231.165244. 9

Cedric Nugteren, Gert-Jan Van den Braak, Henk Corporaal, and Henri Bal. A detailed GPU cache model based on reuse distance theory. In *Proc. of the IEEE International Symposium on High-Performance Computer Architecture (HPCA)*, pages 37–48, 2014. DOI: 10.1109/h-pca.2014.6835955. 79

NVIDIA's Next Generation CUDA Compute Architecture: Fermi. NVIDIA, 2009. 16, 46

Nvidia. *NVIDIA tesla V100 GPU architecture.* 2017. 27, 31

NVIDIA Corp. Pascal l1 cache. https://devtalk.nvidia.com/default/topic/1006066/pasc al-l1-cache/?offset=6 70

NVIDIA Corp. Inside volta: The world's most advanced data center GPU. `https://devblogs` `.nvidia.com/parallelforall/inside-volta/`, May 2017. 1, 17, 26

NVIDIA Corporation. *NVIDIA's Next Generation CUDA Compute Architecture: Kepler TM GK110*, a. 13

NVIDIA Corporation. *NVIDIA GeForce GTX 680*, b. 16

NVIDIA Corporation. *CUDA Binary Utilities*, c. 16

Parallel Thread Execution ISA (Version 6.1). NVIDIA Corporation, CUDA Toolkit 9.1 ed., November 2017. 14

Lars Nyland et al. United States Patent #8,086,806: Systems and Methods for Coalescing Memory Accesses of Parallel Threads (Assignee: NVIDIA Corp.), December 2011. 71

Marek Olszewski, Jeremy Cutler, and J. Gregory Steffan. JudoSTM: A dynamic binary-rewriting approach to software transactional memory. In *Proc. of the ACM/IEEE International Conference on Parallel Architecture and Compilation Techniques (PACT)*, pages 365–375, 2007. DOI: 10.1109/pact.2007.4336226. 96

Jason Jong Kyu Park, Yongjun Park, and Scott Mahlke. Chimera: Collaborative preemption for multitasking on a shared GPU. In *Proc. of the ACM Architectural Support for Programming Languages and Operating Systems (ASPLOS)*, 2015. DOI: 10.1145/2694344.2694346. 92

David A. Patterson and John L. Hennessy. *Computer Organization and Design: The Hardware/-Software Interface*. 2013. 78

Gennady Pekhimenko, Vivek Seshadri, Onur Mutlu, Phillip B. Gibbons, Michael A. Kozuch, and Todd C. Mowry. Base-delta-immediate compression: Practical data compression for on-chip caches. In *Proc. of the ACM/IEEE International Conference on Parallel Architecture and Compilation Techniques (PACT)*, 2012. DOI: 10.1145/2370816.2370870. 60

J. Power, A. Basu, J. Gu, S. Puthoor, B. M. Beckmann, M. D. Hill, S. K. Reinhardt, and D. A. Wood. Heterogeneous system coherence for integrated CPU-GPU systems. In *Proc. of the ACM/IEEE International Symposium on Microarchitecture (MICRO)*, December 2013a. DOI: 10.1145/2540708.2540747. 100

Jason Power, Arkaprava Basu, Junli Gu, Sooraj Puthoor, Bradford M. Beckmann, Mark D. Hill, Steven K. Reinhardt, and David A. Wood. Heterogeneous system coherence for integrated CPU-GPU systems. In *Proc. of the ACM/IEEE International Symposium on Microarchitecture (MICRO)*, pages 457–467, 2013b. DOI: 10.1145/2540708.2540747. 4

M. K. Qureshi and Y. N. Patt.   Utility-based cache partitioning: A low-overhead, high-performance, runtime mechanism to partition shared caches. In *Proc. of the ACM/IEEE International Symposium on Microarchitecture (MICRO)*, pages 423–432, 2006. DOI: 10.1109/micro.2006.49. 101

Xiaowei Ren and Mieszko Lis. Efficient sequential consistency in GPUs via relativistic cache coherence. In *Proc. of the IEEE International Symposium on High-Performance Computer Architecture (HPCA)*, pages 625–636, 2017. DOI: 10.1109/hpca.2017.40. 72

Minsoo Rhu and Mattan Erez.   CAPRI: Prediction of compaction-adequacy for handling control-divergence in GPGPU architectures. In *Proc. of the ACM/IEEE International Symposium on Computer Architecture (ISCA)*, pages 61–71, 2012. DOI: 10.1109/isca.2012.6237006. 44

Minsoo Rhu and Mattan Erez. The dual-path execution model for efficient GPU control flow. In *Proc. of the IEEE International Symposium on High-Performance Computer Architecture (HPCA)*, pages 591–602, 2013a. DOI: 10.1109/hpca.2013.6522352. 49

Minsoo Rhu and Mattan Erez. Maximizing SIMD resource utilization in GPGPUs with SIMD lane permutation. In *Proc. of the ACM/IEEE International Symposium on Computer Architecture (ISCA)*, 2013b. DOI: 10.1145/2485922.2485953. 46

Scott Rixner, William J. Dally, Ujval J. Kapasi, Peter Mattson, and John D. Owens. Memory access scheduling. In *Proc. of the ACM/IEEE International Symposium on Computer Architecture (ISCA)*, pages 128–138, 2000. DOI: 10.1109/isca.2000.854384. 77

James Roberts et al. United States Patent #8,234,478: Using Data Cache Array as a Dram Load/Store Buffer, July 2012. 75

Timothy G. Rogers, Mike O'Connor, and Tor M. Aamodt.   Cache-conscious wavefront scheduling. In *Proc. of the ACM/IEEE International Symposium on Microarchitecture (MICRO)*, 2012. DOI: 10.1109/micro.2012.16. 33, 78, 79, 88

Timothy G. Rogers, Mike O'Connor, and Tor M. Aamodt. Divergence-aware warp scheduling. In *Proc. of the ACM/IEEE International Symposium on Microarchitecture (MICRO)*, 2013. DOI: 10.1145/2540708.2540718. 79, 90

Timothy G. Rogers, Daniel R. Johnson, Mike O'Connor, and Stephen W. Keckler. A variable warp size architecture. In *Proc. of the ACM/IEEE International Symposium on Computer Architecture (ISCA)*, 2015. DOI: 10.1145/2749469.2750410. 53

Sangmin Seo, Gangwon Jo, and Jaejin Lee. Performance characterization of the NAS parallel benchmarks in OpenCL. In *Proc. of the IEEE International Symposium on Workload Characterization (IISWC)*, pages 137–148, 2011. DOI: 10.1109/iiswc.2011.6114174. 10

A. Sethia, D. A. Jamshidi, and S. Mahlke. Mascar: Speeding up GPU warps by reducing memory pitstops. In *Proc. of the IEEE International Symposium on High-Performance Computer Architecture (HPCA)*, pages 174–185, 2015. DOI: 10.1109/hpca.2015.7056031. 91, 92

Ankit Sethia and Scott Mahlke. Equalizer: Dynamic tuning of GPU resources for efficient execution. In *Proc. of the ACM/IEEE International Symposium on Microarchitecture (MICRO)*, 2014. DOI: 10.1109/micro.2014.16. 86

Ryan Shrout. AMD ATI radeon HD 2900 XT review: R600 arrives. *PC Perspective*, May 2007. 75

Mark Silberstein, Bryan Ford, Idit Keidar, and Emmett Witchel. GPUfs: Integrating a file system with GPUs. In *Proc. of the ACM Architectural Support for Programming Languages and Operating Systems (ASPLOS)*, pages 485–498, 2013. DOI: 10.1145/2451116.2451169. 2

Inderpreet Singh, Arrvindh Shriraman, Wilson W. L. Fung, Mike O'Connor, and Tor M. Aamodt. Cache coherence for GPU architectures. In *Proc. of the IEEE International Symposium on High-Performance Computer Architecture (HPCA)*, pages 578–590, 2013. DOI: 10.1109/hpca.2013.6522351. 72

Michael F. Spear, Virendra J. Marathe, William N. Scherer, and Michael L. Scott. Conflict detection and validation strategies for software transactional memory. In *Proc. of the EATCS International Symposium on Distributed Computing*, pages 179–193, Springer-Verlag, 2006. DOI: 10.1007/11864219_13. 99

Michael F. Spear, Maged M. Michael, and Christoph Von Praun. RingSTM: Scalable transactions with a single atomic instruction. In *Proc. of the ACM Symposium on Parallel Algorithms and Architectures (SPAA)*, pages 275–284, 2008. DOI: 10.1145/1378533.1378583. 97

Michael Steffen and Joseph Zambreno. Improving SIMT efficiency of global rendering algorithms with architectural support for dynamic micro-kernels. In *Proc. of the ACM/IEEE International Symposium on Microarchitecture (MICRO)*, pages 237–248, 2010. DOI: 10.1109/micro.2010.45. 45

Ivan E. Sutherland. *Sketchpad a Man-machine Graphical Communication System*. Ph.D. thesis, 1963. DOI: 10.1145/62882.62943. 6

David Tarjan and Kevin Skadron. On demand register allocation and deallocation for a multi-threaded processor, June 30, 2011. U.S. Patent App. 12/649,238. 64

Sean J. Treichler et al. United States Patent #9,098,383: Consolidated Crossbar that Supports a Multitude of Traffic Types, August 2015. 75

Dean M. Tullsen, Susan J. Eggers, Joel S. Emer, Henry M. Levy, Jack L. Lo, and Rebecca L. Stamm. Exploiting choice: Instruction fetch and issue on an implementable simultaneous multithreading processor. In *Proc. of the ACM/IEEE International Symposium on Computer Architecture (ISCA)*, 1996. DOI: 10.1145/232973.232993. 88

Rafael Ubal, Byunghyun Jang, Perhaad Mistry, Dana Schaa, and David Kaeli. Multi2Sim: A simulation framework for CPU-GPU computing. In *Proc. of the ACM/IEEE International Conference on Parallel Architecture and Compilation Techniques (PACT)*, pages 335–344, 2012. DOI: 10.1145/2370816.2370865. 17

Aniruddha S. Vaidya, Anahita Shayesteh, Dong Hyuk Woo, Roy Saharoy, and Mani Azimi. SIMD divergence optimization through intra-warp compaction. In *Proc. of the ACM/IEEE International Symposium on Computer Architecture (ISCA)*, pages 368–379, 2013. DOI: 10.1145/2485922.2485954. 44, 46

Wladimir J. van der Lann. Decuda. http://wiki.github.com/laanwj/decuda/ 14

Jin Wang and Sudhakar Yalamanchili. Characterization and analysis of dynamic parallelism in unstructured GPU applications. In *Proc. of the IEEE International Symposium on Workload Characterization (IISWC)*, pages 51–60, 2014. DOI: 10.1109/iiswc.2014.6983039. 95

Jin Wang, Norm Rubin, Albert Sidelnik, and Sudhakar Yalamanchili. Dynamic thread block launch: A lightweight execution mechanism to support irregular applications on GPUs. In *Proc. of the ACM/IEEE International Symposium on Computer Architecture (ISCA)*, pages 528–540, 2016a. DOI: 10.1145/2749469.2750393. 95

Jin Wang, Norm Rubin, Albert Sidelnik, and Sudhakar Yalamanchili. Laperm: Locality aware scheduler for dynamic parallelism on GPUs. In *Proc. of the ACM/IEEE International Symposium on Computer Architecture (ISCA)*, 2016b. DOI: 10.1109/isca.2016.57. 96

Kai Wang and Calvin Lin. Decoupled affine computation for SIMT GPUs. In *Proc. of the ACM/IEEE International Symposium on Computer Architecture (ISCA)*, 2017. DOI: 10.1145/3079856.3080205. 58, 61, 62

D. Wong, N. S. Kim, and M. Annavaram. Approximating warps with intra-warp operand value similarity. In *Proc. of the IEEE International Symposium on High-Performance Computer Architecture (HPCA)*, 2016. DOI: 10.1109/hpca.2016.7446063. 60, 61

X. Xie, Y. Liang, Y. Wang, G. Sun, and T. Wang. Coordinated static and dynamic cache bypassing for GPUs. In *Proc. of the IEEE International Symposium on High-Performance Computer Architecture (HPCA)*, 2015. DOI: 10.1109/hpca.2015.7056023. 81

Yunlong Xu, Rui Wang, Nilanjan Goswami, Tao Li, Lan Gao, and Depei Qian. Software transactional memory for GPU architectures. In *Proc. of the ACM/IEEE International Symposium on Code Generation and Optimization (CGO)*, pages 1:1–1:10, 2014. DOI: 10.1145/2581122.2544139. 99

Y. Yang, P. Xiang, M. Mantor, N. Rubin, L. Hsu, Q. Dong, and H. Zhou. A case for a flexible scalar unit in SIMT architecture. In *Proc. of the IEEE International Parallel and Distributed Processing Symposium (IPDPS)*, 2014. DOI: 10.1109/ipdps.2014.21. 47

George L. Yuan, Ali Bakhoda, and Tor M. Aamodt. Complexity effective memory access scheduling for many-core accelerator architectures. In *Proc. of the ACM/IEEE International Symposium on Microarchitecture (MICRO)*, pages 34–44, 2009. DOI: 10.1145/1669112.1669119. 77

Hou Yunqing. Assembler for NVIDIA FERMI. `https://github.com/hyqneuron/asfermi` 16

Eddy Z. Zhang, Yunlian Jiang, Ziyu Guo, and Xipeng Shen. Streamlining GPU applications on the fly: Thread divergence elimination through runtime thread-data remapping. In *Proc. of the ACM International Conference on Supercomputing (ICS)*, pages 115–126, 2010. DOI: 10.1145/1810085.1810104. 45

William K. Zuravleff and Timothy Robinson. U.S. patent #5,630,096: Controller for a synchronous dram that maximizes throughput by allowing memory requests and commands to be issued out of order, May 13, 1997. 77

Authors' Biographies

TOR M. AAMODT

Tor M. Aamodt is a Professor in the Department of Electrical and Computer Engineering at the University of British Columbia, where he has been a faculty member since 2006. His current research focuses on the architecture of general-purpose GPUs and energy-efficient computing, most recently including accelerators for machine learning. Along with students in his research group, he developed the widely used GPGPU-Sim simulator. Three of his papers have been selected as "Top Picks" by *IEEE Micro Magazine*, a fourth was selected as a "Top Picks" honorable mention. One of his papers was also selected as a "Research Highlight" in *Communications of the ACM*. He is in the MICRO Hall of Fame. He served as an Associate Editor for *IEEE Computer Architecture Letters* from 2012–2015 and the *International Journal of High Performance Computing Applications* from 2012–2016, was Program Chair for ISPASS 2013, General Chair for ISPASS 2014, and has served on numerous program committees. He was a Visiting Associate Professor in the Computer Science Department at Stanford University from 2012–2013. He was awarded an NVIDIA Academic Partnership Award in 2010, a NSERC Discovery Accelerator for 2016–2019, and a 2016 Google Faculty Research Award.

Tor received his BASc (in Engineering Science), MASc, and Ph.D. at the University of Toronto. Much of his Ph.D. work was done while he was an intern at Intel's Microarchitecture Research Lab. Subsequently, he worked at NVIDIA on the memory system architecture ("framebuffer") of GeForce 8 Series GPU—the first NVIDIA GPU to support CUDA.

Tor is registered as a Professional Engineer in the province of British Columbia.

WILSON WAI LUN FUNG

Wilson Wai Lun Fung is an architect in Advanced Computing Lab (ACL) as part of Samsung Austin R & D Center (SARC) at Samsung Electronics, where he contributes to the development of a next generation GPU IP. He is interested in both theoretical and practical aspects of computer architecture. Wilson is a winner of the NVIDIA Graduate Fellowship, the NSERC Postgraduate Scholarship, and the NSERC Canada Graduate Scholarship. Wilson was one of the main contributors to the widely used GPGPU-Sim simulator. Two of his papers were selected as a "Top Pick" from computer architecture by *IEEE Micro Magazine*. Wilson received his BASc (in Computer Engineering), MASc, and Ph.D. at the University of British Columbia. During his Ph.D., Wilson interned at NVIDIA.

TIMOTHY G. ROGERS

Timothy G. Rogers is an Assistant Professor in the Electrical and Computer Engineering department at Purdue University, where his research focuses on massively multithreaded processor design. He is interested in exploring computer systems and architectures that improve both programmer productivity and energy efficiency. Timothy is a winner of the NVIDIA Graduate Fellowship and the NSERC Alexander Graham Bell Canada Graduate Scholarship. His work has been selected as a "Top Pick" from computer architecture by *IEEE Micro Magazine* and as a "Research Highlight" in *Communications of the ACM*. During his Ph.D., Timothy interned at NVIDIA Research and AMD Research. Prior to attending graduate school, Timothy worked as a software engineer at Electronic Arts and received his BEng in Electrical Engineering from McGill University.